Nelson Advanced Science

D0525857

Exchange and Transport, Energy and Ecosystems

revised edition

John Adds • Erica Larkcom • Ruth Miller

Series Editor: Martin Furness-Smith

Endorsed by

First published in 2000 by:
Nelson Thornes (Publishers) Ltd

Nelson Thornes Ltd
Delta Place
27 Bath Road
CHELTENHAM
GL53 7TH
United Kingdom

This edition published 2004

04 05 06 07 / 10 9 8 7 6 5 4 3 2 1

A catalogue record for this book is available from the British Library

Illustrations by Hardlines and Wearset Ltd

ISBN 0 7487 7487 4

Typeset by Hardlines and Wearset Ltd

Printed in Croatia by Zrinski

Contents

CONTENTS

Introduction

This series has been written by an experienced team of Examiners and others involved directly with the Edexcel Advanced Subsidiary (AS) and Advanced (A) GCE Biology and Biology (Human) specification and its assessment.

Exchange and Transport, Energy and Ecosystems is one of four books in the Nelson Advanced Science (NAS) series. These books have been developed to match the requirements of the Edexcel specification, but they will also be useful for other Advanced Subsidiary (AS) and Advanced (A) courses.

Exchange and Transport, Energy and Ecosystems covers Units 2B, 2H and 3 of the Edexcel specification for the Advanced Subsidiary (AS) course.

A coloured, vertical stripe in the margin indicates that the material is only relevant to Unit 2B (blue stripe) or Unit 2H (pink stripe).

Unit 2B – Exchange, transport and reproduction – for the Biology specification, covers
- exchanges with the environment, including gas exchange, digestion and absorption
- the transport of materials within organisms
- adaptations to the environment
- sexual reproduction.

In all these topics, consideration is given to both plants and animals. The section on adaptations to the environment leads to an understanding that species are adapted to survive in particular environmental conditions, with the emphasis on structural adaptations associated with the presence or absence of water and to the varying oxygen concentrations found in freshwater habitats.

Unit 2H – Exchange, transport and reproduction in humans – for the Biology (Human) specification, covers
- exchanges with the environment, including gas exchange in humans, digestion and absorption
- the transport of materials in humans
- adaptations of humans to extremes of temperature and high altitude
- human reproduction and development.

Unit 3 is common to both Biology and Biology (Human) and covers
- modes of nutrition
- ecosystems
- energy flow
- recycling of nutrients
- energy resources
- human influences on the environment.

This unit also has a practical assessment component requiring candidates to present an individual investigation.

Other resources in this series

NAS *Tools, Techniques and Assessment in Biology* is a course guide for students and teachers. For use alongside the four student texts, it offers ideas and support for practical work, fieldwork and statistics. Key Skills opportunities are identified throughout. This course guide also provides advice on the preparation for assessment tests (examinations).

NAS *Make the Grade in AS Biology with Human Biology* and *Make the Grade in A2 Biology with Human Biology* are Revision Guides for students and can be used in conjunction with the other books in this series. They help students to develop strategies for learning and revision, to check their knowledge and understanding, and to practise the skills required for tackling assessment questions.

Features used in this book – notes to students

The NAS Biology student books are specifically written to help you understand and learn the information provided, and to help you to apply this information to your coursework.

The **text** offers complete and self-contained coverage of all the topics in each Unit. Key words are indicated in bold. The headings for sub-sections have been chosen to link with the wording of the specification wherever possible.

In the margins of the pages, you will find:
- **definition boxes** where key terms are defined. These reinforce and sometimes expand definitions of key terms used in the text.
- **questions** to test your understanding of the topics as you study them. Sometimes these questions take the topic a little further and stimulate you to think about how your knowledge can be applied.

Included in the text are boxes with:
- **background information** designed to provide material which could be helpful in improving your understanding of a topic. This material could provide a link between knowledge gained from GCSE and what you are required to know for AS and A GCE. It could be more information about a related topic or a reminder of material studied at a different level.
- **additional** or **extension** material which takes the topic further. This material is not strictly part of the Edexcel specification and you will not be examined on it, but it can help you to gain a deeper understanding, extending your knowledge of the topic.

In the specification, reference is made to the ability to recognise and identify the general formulae and structure of biological molecules. You will see that we have included the structural chemical formulae of many compounds where we think that this is helpful in gaining an understanding of the composition of the molecules and appreciating how bond formation between monomers results in the formation of polymers. It should be understood that you will not be expected to memorise or reproduce these structural chemical formulae, but you should be able to recognise and reproduce the general formulae for all the molecules specified.

The chapters correspond to the sections of the specification. At the end of Chapter 5, you will find the **practical investigations** linked to the topics covered. These practical investigations are part of the specification and you could be asked questions on them in the Unit tests. Each practical has an introduction, putting it into the context of the topic, and sufficient information about materials and procedure to enable you to carry out the investigation. In addition, there are suggestions as to how you should present your results and questions to help you with the discussion of your findings. In some cases, there are suggestions as to how you could extend the investigation so that it would be suitable as an individual study.

At the end of the book, there are **assessment questions**. These have been selected from past examination papers and chosen to give you as wide a range of different types of questions as possible. These should enable you to become familiar with the format of the Unit Tests and help you to develop the skills required in the examination. **Mark schemes** for these questions are provided so that you can check your answers and assess your understanding of each topic.

Note to teachers on safety

When practical instructions have been given we have attempted to indicate hazardous substances and operations by using standard symbols and recommending appropriate precautions. Nevertheless teachers should be aware of their obligations under the Health and Safety at Work Act, Control of Substances Hazardous to Health (COSHH) Regulations, and the Management of Health and Safety at Work Regulations. In this respect they should follow the requirements of their employers at all times. In particular, they should consult their employer's risk assessments (usually model risk assessments in a standard safety publication) before carrying out any hazardous procedure or using hazardous substances or microorganisms.

In carrying out practical work, students should be encouraged to carry out their own risk assessments, that is, they should identify hazards and suitable ways of reducing the risks from them. However they must be checked by the teacher. Students must also know what to do in an emergency, such as a fire.

Teachers should be familiar and up to date with current advice on safety, which is available from professional bodies.

Teachers are strongly advised to refer to Safety Codes of Practice and Guidelines produced by Education Authorities or by Governing Bodies of schools and colleges. This is particularly important in practical work on students, such as investigations into the effects of exercise, in which students should be sufficiently fit and willing to participate. Some practical activities, in particular those involving measurement of body fat and comparisons of fitness, should be approached with sensitivity and understanding.

Acknowledgements

The authors would like to thank Sue Howarth and David Hartley for their help and support during the production of this book, as well as John Schollar and Dean Madden, The National Centre for Biotechnology Education, The British Nutrition Foundation and Sainsburys plc.

The authors and publishers are grateful for permission to include the following copyright material:

Photographs

Biophoto Associates: 5.5a, b, c, d, e;

Bryan & Cherry Alexander: 4.1 top;

Corbis: Bettmann 4.1 bottom, David Forman/Eye Ubiquitous: 9.17;

Corel 587 (NT): 8.7a;

Dudley Christian: Rothamsted Research 8.10, 8.11;

Erica Larkcom: 4.9, 4.15 left, centre, 8.1 top, bottom, 8.5a, b, 8.6a, b, 8.9, 8.13, 8.18, 9.1a left, right, 9.1b left, right, 9.2 left, right, 9.3, 9.5a, b, c, d, 9.6a, b, 9.21;

FAO: 15410 R Faidutti 9.6c, 12925 M Roodkowsky 9.6d;

Fibrowatt Ltd: 8.12;

Frank Spooner Pictures: Gamma 4.2;

Getty Images Stone: Lorne Resnick: 4.8, James Balog 4.11, Jess Scott 4.14, David Levy 4.15;

John Adds: 1.4a, b, c, 1.7, 1.9c, 2.7b, 2.20, 3.1, 3.2, 5.14a, 5.16, 7.1, P3, P7, page 129

John Bebbington: page1;

Judith Pater: 4.24;

NASA: page 211;

Natural Visions: Heather Angel: 6.1a, b, c;

Ruth Miller: 5.11, 7.14;

Science Photolibrary: 2.16, Department of Clinical Radiology/Salisbury District Hospital 1.17a, Manfred Kage 1.17b, Eric Grave 1.26b, Dr Jeremy Burgess 2.12a, J C Revy 2.12b, Astrid/Hanns-Frieder Michler 3.8a, Science Pictures Ltd 3.8b, Biophoto Associates 4.12, 6.6c, Andrew Syred 5.1, 5.8a, David Scharf 5.8a, Simon Fraser/Northumbrian Environmental Management Ltd 8.16;

Telegraph Colour Library: cover.

Artwork

Figure 2.14 from Salisbury & Ross, *Plant Physiology*, 3rd edn, p.119, fig 6–8 (ISBN 0534044824), Wadsworth.

Figure 2.18 from Griffin & Redmore (1993), *Human Systems*, 3rd edition, p.35, fig 3.2b (ISBN 0534044824), Nelson.

About the authors

John Adds is Chief Examiner for AS and A GCE Biology and Biology (Human) and Head of Biology at Abbey College, London.

Erica Larkcom is Deputy Director of Science and Plants for Schools at Homerton College, Cambridge, and a former Subject Officer for A level Biology.

Ruth Miller is a former Chief Examiner for AS and A GCE Biology and Biology (Human) and former Head of Biology at Sir William Perkins's School, Chertsey.

Unit 2 – What do I need to study?

Topic	Page no.	Unit 2B*	Unit 2H*
Chapter 1			
Exchange processes	2	✓	✓
The nature of exchange surfaces	3	✓	✓
Histology of epithelia	4		✓
Respiratory gas exchange	5	✓	✓
Gas exchange in simple animals	5	✓	
Gas exchange in flowering plants	6	✓	
Gas exchange in humans	7	✓	✓
Lung capacities	10	✓	✓
Control of breathing	11	✓	✓
Effects of smoking	14		✓
Digestion and absorption	15	✓	✓
Chapter 2			
The need for transport systems	24	✓	✓
Transport in flowering plants	26	✓	
Transport in humans and other mammals	39	✓	✓
Structure and function of the heart	39	✓	✓
Control of heart beat	42	✓	✓
The electrocardiogram	43		✓
Pacemakers	43		✓
Arteries, capillaries and veins	45	✓	✓
Blood and body fluids	45	✓	✓
Interchange of materials	49	✓	✓
Chapter 3			
Adaptations to the environment	51	✓	
Chapter 4			
Human ecology	58		✓
Chapter 5			
Sexual reproduction	86	✓	✓
Meiosis	86	✓	✓
Reproduction in flowering plants	93	✓	
Reproduction in humans	99	✓	✓
Fertilisation, implantation and early development	106	✓	✓
Birth and lactation	109	✓	✓
Growth and physical development	110		✓

*2B = Biology Unit 2; 2H = Biology (Human) Unit 2

Exchange, Transport, Adaptation and Reproduction

Large skipper (Ochlodes venata) *feeding on a flower of the fragrant orchid* (Gymnadenia conopsea). *"Pollinia" (pollen) of the orchid can be seen on the "elbow" of the proboscis. (Photograph taken near Dorking, Surrey, UK.)*

1 Exchanges with the environment

All living organisms constantly need to exchange materials with the environment in order to survive and grow. Exchange of materials involves:
- **respiratory gases** – all organisms need to respire to release energy from their food. Most take up oxygen from the environment and, as a consequence of the catabolic reactions involved, release carbon dioxide into the environment as a waste product.
- **nutrients** – a supply of nutrients is required by all living organisms in order to provide energy and materials for growth and reproduction. Organisms differ in the ways in which these nutrients are acquired. **Heterotrophic** organisms, such as animals, obtain ready-made sources of food from their environment. This food consists of complex organic (carbon-containing) compounds which need to be digested into soluble molecules before they can be absorbed and used. **Autotrophic** organisms, such as green plants, use light energy to build up organic molecules from carbon dioxide, water and mineral ions in the process of photosynthesis.
- **excretory products** – all organisms produce waste products as a result of their metabolic processes. In heterotrophic organisms, carbon dioxide is produced as a waste product of respiration, together with nitrogenous waste substances. These nitrogenous substances could become toxic if allowed to accumulate, so they are excreted together with any other compounds in excess of requirements. In autotrophic organisms, the main excretory products are carbon dioxide from respiration and oxygen from photosynthesis.

Exchange processes

The physical processes involved with exchange processes at cellular level are:
- **passive** – such as diffusion in the exchange of gases associated with the processes of respiration and photosynthesis in the leaves of flowering plants, and respiration across the respiratory surfaces of animals
- **active** – as in the uptake of mineral ions against concentration gradients in the roots of flowering plants.

These active and passive processes have been described in *Molecules and Cells*, but it is useful here to consider the features of exchange surfaces which affect the rate at which the exchange processes occur. The rate of diffusion across a membrane depends on:
- the diffusion (or concentration) gradient: the steeper the gradient, the faster the rate of diffusion
- the size of the diffusing molecules: smaller molecules such as oxygen and carbon dioxide diffuse more rapidly than larger molecules such as glucose
- the area over which it occurs: the greater the area, the more rapid the diffusion
- the distance over which it occurs: the shorter the distance, the more rapid the rate

- the nature of any barrier through which the molecules must pass: membranes need to be permeable to the diffusing molecules.

The area of any exchange surface must be large enough to enable the process to occur efficiently. In small organisms, such as the single-celled organism *Amoeba*, exchanges with the environment occur across the entire body surface through the cell surface membrane. In larger, multicellular organisms, an increase in volume results in a decrease in the surface area: volume ratio. In other words, there is less surface area per unit volume of organism and the area over which exchanges take place may not be large enough to satisfy the needs of the organism. It is easy to demonstrate that as the surface area of a cube increases, the surface area:volume ratio decreases (Figure 1.1).

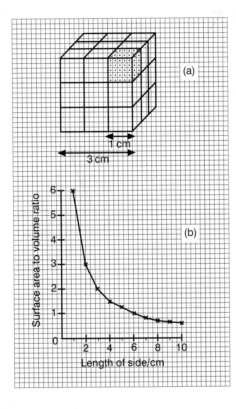

Figure 1.1 How surface area:volume ratio changes with size: (a) The small (dotted) cube has sides of length 1 cm. It has six faces, so its surface area = 1 × 1 × 6 = 6 cm². Its volume = 1 × 1 × 1 = 1 cm³. The surface area:volume ratio is therefore 6:1. The large cube has sides of length 3 cm. Its surface area is 3 × 3 × 6 = 54 cm², and its volume is 3 × 3 × 3 cm = 27 cm³. The surface area:volume ratio of this cube is therefore 54:27 or 2:1. As the length of side of a cube increases, the ratio of the surface area to volume decreases, i.e. there is a smaller surface area per unit of volume. This is illustrated by the graph in (b)

As organisms increase in size and there is a corresponding decrease in the surface area to volume ratio, there is an increase in bulk of the organism. This means that the distance over which diffusion occurs becomes greater and consequently the rate of diffusion becomes slower.

Features of exchange surfaces

Exchange surfaces are the sites where materials are exchanged between the organism and the environment. In simple organisms, this process occurs over the entire surface but in more complex, multicellular organisms there are specialised regions adapted for a particular function. Most of the exchanges between flowering plants and their environment occur through the roots or through the aerial parts, particularly the leaves. In mammals, most exchanges occur internally and involve **epithelial tissues**.

BACKGROUND

Epithelial tissues are found on the internal and external surfaces of animal organs and may have several roles, depending on their location. Many epithelia protect underlying tissues against water loss, damage (by rubbing), pressure or infection. In addition, epithelial tissues may be involved in processes such as respiratory gas exchange, the uptake or release of nutrients and excretion.

A **simple epithelium** consists of cells arranged in a single layer, whereas **compound** or **stratified epithelia** are composed of several layers of cells. The compound epithelia, being thicker, often form impervious barriers on the external surface, but the simple epithelia form efficient exchange surfaces.

The main features of simple epithelial tissues are that:
- they form continuous layers on internal and external surfaces
- the cells are held together by a thin layer of intercellular substance containing hyaluronic acid which sticks them together
- the cells rest on a basement membrane made up largely of collagen fibres
- there are no blood cells present
- the free surfaces of the cells may be highly specialised
- damaged cells are rapidly replaced by cell division.

Types of simple epithelia

Cuboidal epithelium (Figure 1.2) is the simplest type of epithelium and consists of cube-shaped cells, each with a centrally-situated spherical nucleus. The cells are closely packed together and appear pentagonal or hexagonal in outline when viewed from above. This type of epithelium occurs in the nephrons of the kidney and lines the salivary and pancreatic ducts. It is also present in many glands (mucus, salivary, sweat and thyroid), where it has a secretory function.

Squamous epithelium (Figure 1.3) consists of thin, flattened cells with little cytoplasm. The nucleus of each cell is disc-shaped and centrally situated. Cytoplasmic connections exist between adjacent cells. The cells fit closely together and, when viewed from above, the margins of the cells are seen to be irregular. This type of epithelium is found in the Bowman's capsule of the kidney, the alveoli of the lungs and lining the blood vessels and the chambers of the heart.

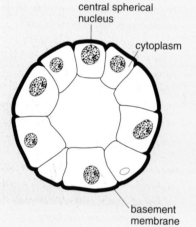

Figure 1.2 Cuboidal epithelium – found in nephrons, salivary and pancreatic ducts and secretory glands

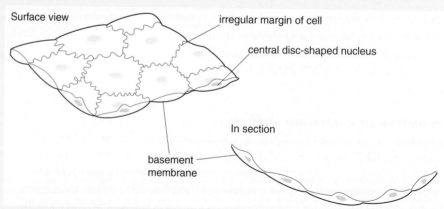

Figure 1.3 Squamous epithelium – found in the Bowman's capsule, alveoli, and in the lining of blood vessels and heart chambers

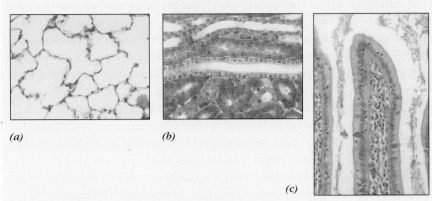

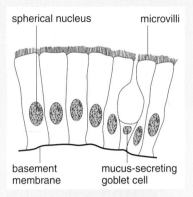

Figure 1.4 The different types of epithelia, as seen using a light microscope: (a) squamous epithelium (alveoli); (b) cuboidal epithelium (kidney tubule); (c) columnar epithelium (ileum)

Figure 1.5 Columnar epithelium – found in the lining of the stomach and intestine, and in some ducts of the kidney

Columnar epithelium (Figure 1.5) is made up of tall, narrow cells. There is a large spherical nucleus near the base of each cell and the free surface often possesses **microvilli**. Mucus-secreting goblet cells are often found amongst the columnar cells. This tissue lines the stomach and intestine, and is also present in some ducts of the kidney.

Features of gas exchange surfaces and the need for ventilation mechanisms

As we have seen, most organisms respire aerobically, and so they take up oxygen and release carbon dioxide.

Respiring cells are constantly using up oxygen and releasing carbon dioxide, so concentration gradients exist between the organism and its environment with respect to these gases. Usually within the organism there is a lower concentration of oxygen and a higher concentration of carbon dioxide than in the environment, so oxygen tends to diffuse in and carbon dioxide diffuses out.

The area of the gas exchange surface must be large enough to provide sufficient oxygen for the organism's requirements. In very small organisms, such as the single-celled *Amoeba*, where the surface area : volume ratio is large, the general body surface is the gas exchange surface. In such organisms, gas exchange takes place through the cell surface membrane, oxygen diffusing in and carbon dioxide diffusing out. With larger, multicellular organisms, an increase in volume results in a decrease in the surface area : volume ratio. There is less surface area per unit volume of organism and exchange of gases through the body surface may not be enough to satisfy the organism's needs. In such cases, specialised gas exchange surfaces exist in the form of lungs or gills, providing a large area over which the exchange can occur.

In small organisms, the distances over which the diffusion of the gases occurs are small, but with increase in size there is a corresponding increase in bulk. This results in an increase in the distance of the respiring cells from the gas exchange surface, slowing the rate of diffusion. In some larger organisms such

> **DEFINITION**
>
> **Microvilli** (singular microvillus) are minute, finger-like projections of the surface membrane of many epithelia. Microvilli are about 0.5 to 1.0 μm in length and are too small to be seen individually with a light microscope. Microvilli increase the surface area of cells for absorption.

> **BACKGROUND INFORMATION**
>
> The features of gas exchange surfaces are determined by the factors which affect the rate of diffusion. We have already seen that the rate of diffusion will depend on the existence of concentration gradients, but other factors that need to be considered are:
> * the area over which diffusion occurs
> * the distance over which diffusion occurs
> * the nature of any barrier through which the molecules must pass
> * the nature of the diffusing molecules.

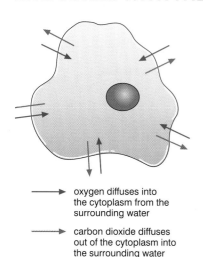

oxygen diffuses into
the cytoplasm from the
surrounding water

carbon dioxide diffuses
out of the cytoplasm into
the surrounding water

Figure 1.6 Gas exchange in Amoeba

as the flatworms, where there is no special gas exchange surface, the body is flattened. This increases the efficiency of diffusion as no respiring cells are far from the gas exchange surface. In different organisms where specialised gas exchange surfaces are present, other mechanisms have evolved which improve the efficiency of gas exchange. A breathing mechanism often exists, bringing fresh supplies of air or water in contact with the gas exchange surface and maintaining a high concentration of oxygen. In addition, the concentration gradients are maintained by an internal transport system, such as the blood circulatory system, which brings deoxygenated blood to the gas exchange surface and removes the oxygenated blood. In such cases, the oxygen-carrying capacity of the blood is increased by the presence of a respiratory pigment. This is a special compound to which oxygen becomes attached. In mammals, this pigment is **haemoglobin** and is present in specialised blood cells called erythrocytes.

The gas exchange surface needs to be permeable to the respiratory gases. All cell surface membranes are permeable to oxygen, carbon dioxide and water. Single-celled organisms, such as *Amoeba* (Figure 1.6), have a large surface area available for gas exchange per unit volume. Gas exchange occurs over the whole of the external surface through the cell surface membrane. Oxygen diffuses into the organism from a high concentration to a lower concentration within the organism. Carbon dioxide produced by respiration diffuses from a higher concentration within the organism to a lower concentration outside. The cell surface membrane is an efficient gas exchange surface.

Gas exchange in flowering plants

Gas exchange in flowering plants involves the aerial parts, mainly the leaves and stems as well as parts of the root. Leaves have a large surface area : volume ratio, which is favourable for the exchange of gases. Oxygen or carbon dioxide reach the

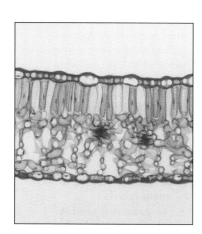

Figure 1.7 Vertical section through part of a leaf blade as seen with a light microscope (low magnification)

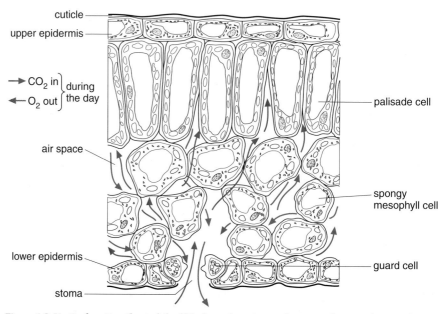

cuticle

upper epidermis

CO_2 in ⎫ during
O_2 out ⎭ the day

air space

palisade cell

spongy mesophyll cell

lower epidermis

guard cell

stoma

Figure 1.8 Vertical section through leaf blade to show gas exchange surfaces and gas exchange during the day, when the rate of photosynthesis is greater than the rate of respiration

respiring cells via the **stomata**, which are pores in the epidermis of the leaves (Figure 1.8). Inside the leaf, the large intercellular air spaces in the spongy mesophyll facilitate the diffusion of gases and the cells bordering these air spaces increase the total area available for gas exchange still further. During daylight, when photosynthesis is occurring, the cells inside the leaf are using carbon dioxide. This produces a concentration gradient of carbon dioxide between the interior of the leaf and the external atmosphere. The rate of diffusion of carbon dioxide is directly proportional to the concentration gradient, but it is also affected by factors such as the number and size of the stomata, the cuticle of the leaf and the layer of air surrounding the leaf. As these are factors which also affect the movement of water in the plant, they will be described in more detail in Chapter 2.

It must be remembered that the situation is slightly different in green plants (Table 1.1). Respiration takes place all the time, so oxygen is continually taken up by respiring cells and carbon dioxide is released. During the hours of daylight, photosynthesis will occur in the palisade and spongy cells in the mesophyll of the leaves, involving the uptake of carbon dioxide and the release of oxygen. As photosynthesis takes place at a more rapid rate than respiration, during the day the concentration gradients of oxygen and carbon dioxide are reversed: carbon dioxide diffuses in and oxygen diffuses out.

Table 1.1 *Gas exchange in flowering plants*

Feature	Day	Night
respiration occurring	✓	✓
photosynthesis occurring	✓	✗
state of stomata	open	closed
concentration of CO_2 in leaf	low	high
concentration of CO_2 in atmosphere	higher	lower
concentration of O_2 in leaf	high	low
concentration of O_2 in atmosphere	lower	higher
net gas exchange	O_2 diffuses out CO_2 diffuses in	CO_2 diffuses out O_2 diffuses in

Gas exchange in humans

The gas exchange system in humans and other mammals consists of two major parts, a **conducting system**, for the conduction of inspired and expired gases, and an **interface** for the exchange of gases between air and blood. The conducting system begins with the nasal passages and continues as the **trachea**. The trachea divides to form the left and right **bronchi** which supply the lungs. The bronchi divide into numerous smaller **bronchioles** which eventually lead into the **alveoli**, where gas exchange occurs (Figure 1.9).

The inside of the thoracic cavity is lined with a **pleural membrane**. Each lung is also surrounded by a separate pleural membrane. The space between the two pleural membranes is filled with a thin film of fluid (known as **pleural fluid**), which prevents friction between them during breathing. This fluid also

EXCHANGES WITH THE ENVIRONMENT

Figure 1.9 Structure of the human respiratory system: (a) the lungs and associated structures in the thorax; (b) the relationship between alveoli and the surrounding blood vessels; (c) a thin section of lung tissue as seen using a light microscope, showing alveoli

(a)

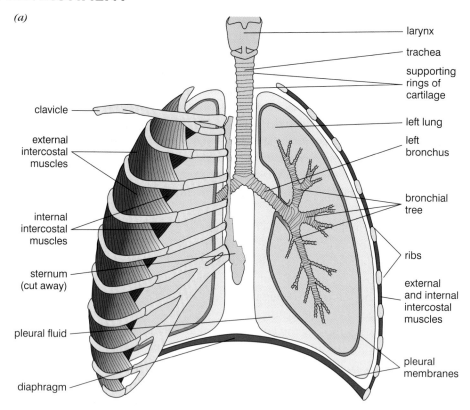

- larynx
- trachea
- supporting rings of cartilage
- left lung
- left bronchus
- bronchial tree
- ribs
- external and internal intercostal muscles
- pleural membranes
- clavicle
- external intercostal muscles
- internal intercostal muscles
- sternum (cut away)
- pleural fluid
- diaphragm

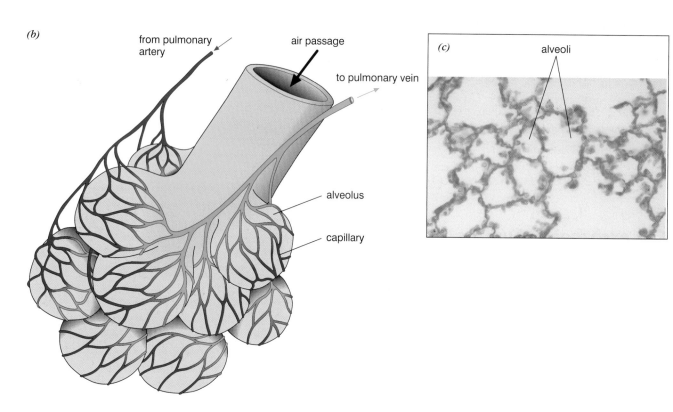

(b)

- from pulmonary artery
- air passage
- to pulmonary vein
- alveolus
- capillary

(c)

- alveoli

causes the outer surface of the lungs to adhere closely to the inside of the thoracic cavity during inspiration.

Alveoli are the sites of gas exchange. Each alveolus is a pocket-shaped structure, 100 to 300 μm in diameter, open on one side, and lined with extremely flattened epithelial cells (Figure 1.9c). It has been estimated that there are about 300 million alveoli in the human lungs, which give a considerable total surface area (40 to 60 m^2) for gas exchange. Each alveolus consists of epithelium and connective tissue (including elastic fibres and collagen) and is surrounded with capillaries. The epithelium cells are of two types, known as **Type I** and **Type II pneumocytes**. Type I pneumocytes are relatively large, extremely flattened cells and make up most of the alveolar wall. Type II pneumocytes secrete **surfactant**, a mixture of lipids and proteins, which helps to prevent the alveolus from collapsing completely and 'sticking shut' as air moves out of the lungs during expiration.

The capillaries surrounding the alveoli are referred to as pulmonary capillaries. They are about 7 to 10 μm in diameter and form a dense network around each alveolus (Figure 1.9b).

The structure of each alveolus, with its surrounding capillaries, is well adapted for the process of gas exchange. Each alveolus is very thin-walled and there are millions of alveoli in each lung. The barrier for diffusion of gases between air in the alveoli and the blood is of minimal thickness, less than 0.5 μm, which increases the efficiency of diffusion of gases.

The lungs have a number of defence mechanisms against inhaled microorganisms and small particles such as carbon in smoke (see page 14). These defence mechanisms include filtration of inhaled air by the nose, the cough reflex and cilia and mucus within the larger airways. Cilia and mucus help to trap small particles in inspired air. Some particles, such as carbon, may reach the alveoli and are engulfed by wandering phagocytic cells known as **alveolar macrophages** or dust cells. These are derived from **monocytes** in blood, which migrate through blood vessel walls into various organs to become macrophages.

Ventilation in humans

Ventilation is the term given to the process of breathing, that is, the movement of air into and out of the gas exchange system. Movement of air into the lungs is referred to as **inspiration**, while the movement of air out of the lungs is **expiration**.

Air moves into, or out of, the lungs as a result of differences in pressure between the atmosphere and air in the alveoli. When the atmospheric pressure (normally about 101 kPa, or 760 mmHg) is greater than the pressure within the lungs, air will tend to flow down this pressure gradient and into the lungs. This is how inspiration occurs. When the pressure in the lungs is greater than atmospheric pressure, air moves out of the lungs and into the atmosphere. The mechanism of ventilation therefore depends on two gas pressure gradients, one in which the pressure within the lungs is lower than atmospheric pressure for inspiration to occur, and one in which the pressure in the lungs is higher than atmospheric pressure for expiration to occur. These pressure gradients are brought about by changes in the volume of the thorax which, in turn, are produced by contraction or relaxation of muscles.

Inspiration

During inspiration (breathing in), the volume of the thorax is increased by the movement of the ribs upwards and outwards, and by contraction of the **diaphragm**. The diaphragm is a sheet of muscle tissue which separates the thoracic and abdominal cavities. As the diaphragm contracts (when it is stimulated by the phrenic nerve), it flattens and moves downwards, which increases the length of the thoracic cavity. Contraction of the **external intercostal muscles** pulls the anterior end of each rib upwards and outwards, increasing the diameter of the thorax. As the overall volume of the thorax increases, the pressure within the lungs decreases and air moves into the lungs.

Expiration

Expiration at rest (referred to as quiet expiration) is brought about mainly by the relaxation of the diaphragm and external intercostal muscles, and contraction of the **internal intercostal muscles**. The changes which occur are essentially the opposite of those described for inspiration, that is, the volume of the thorax decreases and, as a result, the pressure within the lungs increases. This produces a pressure gradient and expiration follows. The lungs are naturally elastic and so tend to deflate as the pressure within them decreases. This is referred to as elastic recoil of the lungs.

The pressure within the lungs therefore varies during inspiration and expiration. At the end of expiration during quiet breathing, the pressure within the lungs is the same as atmospheric pressure. As inspiration starts, the pressure within the lungs drops to about 0.4 kPa below atmospheric pressure. During quiet expiration, the pressure within the lungs initially increases to about 0.4 kPa above atmospheric pressure, but returns to the atmospheric pressure value by the time quiet expiration is completed.

The diaphragm and intercostal muscles are not the only muscles which are involved in ventilation. During quiet breathing, movement of the diaphragm accounts for about 75 per cent of the volume of the air breathed. However, during forced inspiratory and expiratory efforts, many other muscles are used, including the abdominal muscles, which contract during forced expiration. Forced expiratory efforts can greatly increase the pressure within the lungs. For example, forced expiration when trying to blow up a balloon may increase the pressure within the lungs by 13 kPa or more.

The mechanisms of inspiration and expiration are summarised in Figures 1.10, 1.11 and 1.12.

Lung capacities

A **spirometer** (Figure 1.13) is a device which is used to measure and record the volumes of air inspired and expired. These volumes are of great importance as they can indicate whether or not adequate ventilation of the

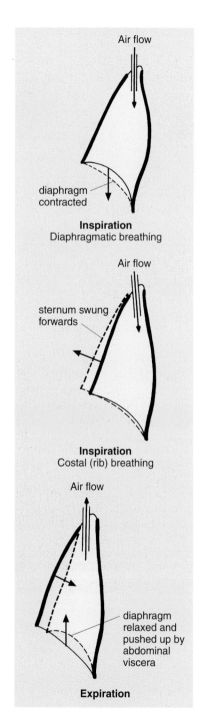

Inspiration
Diaphragmatic breathing

Air flow

diaphragm contracted

Air flow

sternum swung forwards

Inspiration
Costal (rib) breathing

Air flow

diaphragm relaxed and pushed up by abdominal viscera

Expiration

Figure 1.10 The mechanics of inspiration and expiration (see also Figures 1.11 and 1.12)

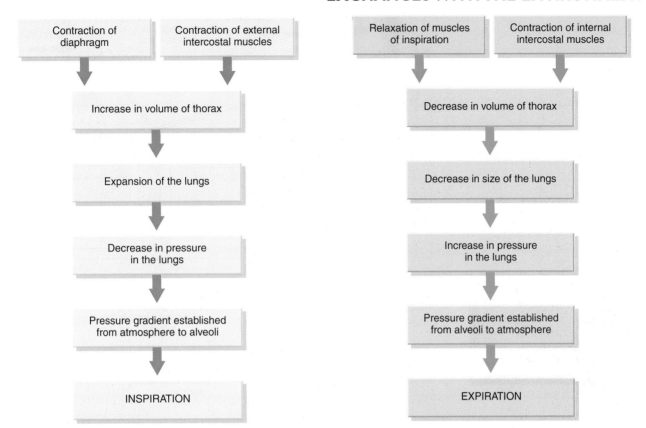

Figure 1.11 *Mechanism of inspiration*

Figure 1.12 *Mechanism of expiration*

lungs is occurring so that there is normal exchange of oxygen and carbon dioxide between alveolar air and the pulmonary capillary blood.

The recording of the volumes of air inspired and expired, usually as a function of time, is referred to as a **spirogram** (Figure 1.14). The volume of air breathed in or out by an adult human during quiet breathing is about 500 cm³. This is referred to as the resting **tidal volume** (TV). After a person has expired tidal air, it is possible to force more air out of the lungs. The maximum volume of air which can be forcibly expired after a tidal expiration is known as the **expiratory reserve volume** (ERV). The **inspiratory reserve volume** (IRV) is the volume of air which can be inspired over and above a tidal inspiration. Even after breathing out as far as possible, air remains in the alveoli. This air, which keeps the alveoli partly inflated and enables gas exchange to continue between breaths, is known as the **residual volume** (RV).

The total of IRV + TV + ERV is known as the **vital capacity** (VC). Vital capacity is related to the body size of a person and is usually about 2.6 dm³ m⁻² body surface area in males and 2.1 dm³ m⁻² in females. It is higher in swimmers and divers, and lower in older people and in people who have diseases of the lungs, such as emphysema. Vital capacity is also affected by posture, being greater when the person is standing upright than when lying down.

Control of breathing
Breathing occurs due to the rhythmic activity of nerves, which send impulses to the diaphragm and intercostal muscles, causing them to contract. Breathing

DEFINITIONS
- **Tidal volume** is the volume of air breathed in or out during quiet breathing.
- **Expiratory reserve volume** (ERV) is the maximum volume of air which can be forcibly expired after a tidal expiration.
- **Inspiratory reserve volume** (IRV) is the volume of air which can be inspired over and above a tidal respiration.

QUESTION
Use Figure 1.14 to explain what is meant by the term **total lung capacity**.

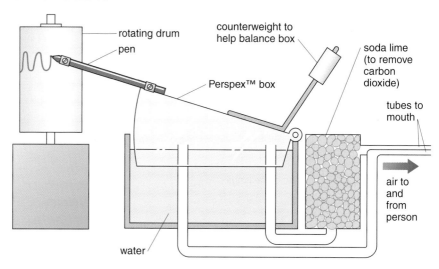

Figure 1.13 A spirometer

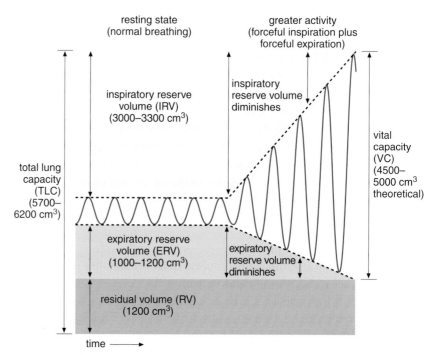

Figure 1.14 Major subdivisions of lung volumes. The red lines show how the tidal volume can change

is controlled by two mechanisms, a voluntary system and an automatic system, both situated in the brain. The voluntary system is used to control breathing in activities such as speaking, singing and playing a wind instrument, and the automatic system regulates breathing to match the metabolic needs of the body.

The basic regular rhythm of breathing seems to be due to an area of the brain known as the **medullary rhythmicity centre**. This is situated in the medulla oblongata, part of the hindbrain. The medullary rhythmicity centre in turn consists of two interconnected areas, the **inspiratory centre** and the **expiratory centre**. Nerve impulses from the inspiratory centre stimulate inspiration, and impulses from the expiratory centre stimulate expiration. The medullary rhythmicity centre is similar to the pacemaker region of the heart

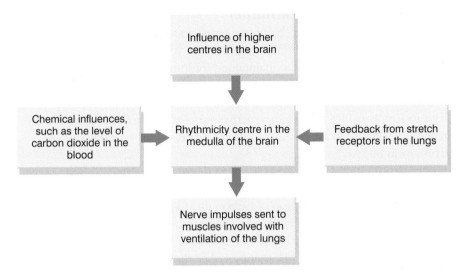

Figure 1.15 Diagram to show some of the factors involved in the control of breathing

(described in Chapter 2) as it has a natural rhythm, which can be modified by external influences.

The activity of the medullary rhythmicity centre is influenced by feedback information that comes from many sensors throughout the nervous system. Stimuli affecting the medullary rhythmicity centre can be divided into two groups: chemical and non-chemical. Chemical stimuli include the respiratory gases (oxygen and carbon dioxide) and changes in the pH of arterial blood. Non-chemical stimuli include feedback from stretch receptors situated in the lungs.

The effects of these types of stimuli are summarised below.

- **Carbon dioxide** Chemoreceptors in the medulla oblongata (Figure 1.16) are sensitive to changes in the carbon dioxide content of arterial blood. A slight increase in carbon dioxide has a stimulating effect, resulting in faster breathing with a greater volume of air moving in and out of the lungs each minute. This increases the removal of carbon dioxide via the lungs and so brings the arterial concentration of carbon dioxide back towards the normal range. Decreased carbon dioxide has the opposite effect, resulting in inhibition of the medullary rhythmicity centre and slower breathing. Breathing may stop entirely for a few seconds if arterial carbon dioxide drops to about 4.6 kPa (the normal range for arterial carbon dioxide is about 5.1 to 5.3 kPa).
- **Oxygen** The precise role of oxygen in controlling breathing is not clear, but a decrease in arterial oxygen below about 9.3 kPa causes stimulation of the inspiratory centre. An increase in arterial oxygen has little effect on breathing.
- **pH of blood** A decrease in arterial blood pH (that is, an increase in acidity) has a stimulating effect on breathing and ventilation increases.
- **Non-chemical stimuli** These include feedback from stretch receptors in the lungs which, as the lungs inflate, send inhibitory impulses to the inspiratory centre. Impulses from sensory nerve endings in joints and tendons help to control breathing during exercise.

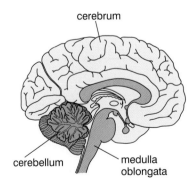

Figure 1.16 Section through the centre of a human brain to show the position of the medulla oblongata

Effects of exercise on ventilation

Exercise can be considered to be of two types, moderate and very severe. These types of exercise have different effects on ventilation. Moderate exercise is a type of exercise that can be maintained for long periods of time, such as walking briskly at 8 km per hour, or steady running. During moderate exercise, ventilation increases steadily in proportion to the extent of the exercise, in order to supply sufficient oxygen to active muscles. The exact reasons for this increase in ventilation are not clear, but are probably due to a number of factors including:

- nervous stimuli from 'higher centres' in the brain and from sensory nerve endings in joints and muscles
- an increase in the production of carbon dioxide
- production of lactic acid by muscles. When muscles receive insufficient oxygen, anaerobic respiration occurs and lactic acid is produced as a result. Lactic acid decreases the pH, that is, makes the tissue slightly acidic.

Table 1.2 shows the ventilation rate of a man walking at different speeds. Ventilation rate is the total volume of air breathed in (or out) per minute.

Table 1.2 *Ventilation rate of a man walking at different speeds*

Walking speed / km per hour	Ventilation rate / dm^3 per minute
rest	10.0
3.2	19.0
4.8	25.0
6.4	37.0
8.0	60.0

QUESTION

Describe the relationship between the ventilation rate and the extent of exercise.

QUESTION

Apart from an increase in pulmonary ventilation, what other physiological changes will occur in a person during and after very severe exercise?

Very severe exercise, such as running 100 m at top speed, can be maintained only for a relatively short period of time. At the end of this exercise, the runner is completely exhausted. Breathing remains much above the resting value for a prolonged period after the exercise is over. As an example, in a man who ran 200 m in 23.4 seconds, it took 27 minutes for his ventilation rate to return to normal. After a 400 m race, followed by vigorous gymnastics, the ventilation rate returned to normal in 44 minutes. In the case of a 100 m sprint, the runner may scarcely draw breath during the race, but will breathe heavily for some time afterwards.

Carbon monoxide, cigarette smoke and gas exchange

Carbon monoxide (CO) is a pollutant gas, which is present in car exhaust fumes and cigarette smoke. Inadequately ventilated central heating boilers may also produce carbon monoxide.

Carbon monoxide combines very readily with haemoglobin, that is, it has a high affinity for haemoglobin and combines approximately 250 times more readily than oxygen. It combines with the haem group of haemoglobin, forming **carboxyhaemoglobin**. This prevents oxygen from combining with the haemoglobin, so reducing the volume of oxygen that can be transported by the blood. Death occurs in humans who are exposed to concentrations of carbon monoxide of around 1000 ppm (parts per million), which corresponds to a blood carboxyhaemoglobin concentration of about 60 per cent. Exposure to lower concentrations of carbon monoxide produces a number of adverse effects, such as dizziness, headaches and mental confusion.

In smokers, the concentration of carboxyhaemoglobin varies between about 1.2 and 9 per cent. This reduces the oxygen carrying capacity of the blood, resulting in breathlessness, and is a factor in the development of heart disease. Women who smoke during pregnancy risk reducing the transfer of oxygen to the developing fetus, which, as a result, grows more slowly than normal. This can result in the birth of an underweight baby.

Cigarette smoke also contains many other harmful substances, including nicotine, aromatic hydrocarbons and phenols. The hydrocarbons include substances that act as tumour initiators, that is, they are able to start a process leading to the development of cancer (Figure 1.17a). In addition to this increased risk of causing lung cancer, cigarette smoke has a number of other harmful effects, including increasing the risk of chronic bronchitis and emphysema.

Cigarette smoke irritates the lining (mucosa) of the trachea and bronchi, increasing the production of mucus and making the mucus thicker. The smoke also paralyses the cilia, which normally move mucus up the trachea towards the back of the throat. As a result, mucus tends to accumulate in the airways (resulting in a condition known as chronic bronchitis), making infection almost inevitable. Bronchitis obstructs airflow and, as a result, there is inefficient gas exchange. This results in laboured breathing and a feeling of breathlessness.

Emphysema is a destructive, incurable, disease of the lungs due to damage to the connective tissue. The alveoli enlarge, their walls rupture and they fuse to form large, dilated, irregular air spaces (Figure 1.17b). This destruction of lung tissue greatly decreases the surface area for gas exchange and, as a result, the blood may be poorly oxygenated. Sufferers from emphysema may find it difficult, for example, to climb a flight of stairs without experiencing fatigue and having to stop to 'catch their breath'. Emphysema can cause serious distress and death from respiratory failure, heart failure or chest infection. There is no cure for emphysema, which is responsible for about 20 000 deaths per year in Great Britain.

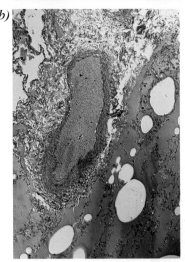

(a)

(b)

Figure 1.17 (a) X-ray of lung cancer caused by smoking cigarettes; (b) part of an emphysematous lung, showing destruction of alveolar walls and the production of large air spaces

Digestion and absorption

Digestion

Digestion is the process in which complex food molecules are broken down into simpler molecules. It can be divided into:
- **mechanical** digestion
- **chemical** digestion.

The structure of the alimentary canal is specialised for the digestion of and the absorption of the products of digestion in heterotrophic organisms.

> **DEFINITION**
> The **lumen** is the cavity of the gut, through which food passes.

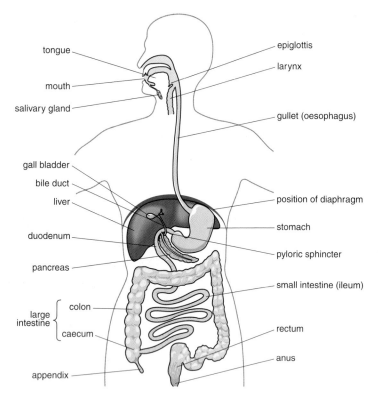

Figure 1.18 Human alimentary canal and associated organs

Structure of the alimentary canal in relation to digestion and absorption

The alimentary canal, or gut, has a basic common structure along its length, although it is specialised in certain regions to carry out its various roles (Figure 1.18). It extends as a tube from the **mouth** to the **anus** and along its length the wall is composed of four layers. These are:

- the **mucosa**, the innermost layer surrounding the **lumen** (Figure 1.19), made up of glandular epithelium (epithelium containing secretory cells) and connective tissue containing blood vessels and lymph vessels
- a layer of connective tissue, the **submucosa**, containing nerves, blood vessels and lymph vessels, together with elastic fibres and collagen
- the **muscularis externa**, composed of circular and longitudinal layers of smooth muscle fibres
- the outermost layer, the **serosa**, made up of loose connective tissue.

Along the whole length of the gut, the glandular epithelium of the mucosa contains **goblet cells** which secrete **mucus**. The mucus lubricates the passage of food along the gut and also protects it from the digestive action of enzymes. In the mucosa of the stomach, there are simple, tubular **gastric glands** which secrete **gastric juice**. In the mucosa of the duodenum and ileum, the **intestinal glands** in the **crypts of Lieberkühn** secrete **intestinal juice** which contains mucus and enterokinase.

The mucosa is separated from the submucosa by a thin layer of smooth muscle, the **muscularis mucosa**. In the submucosa of the duodenum, **Brunner's glands** produce an alkaline solution containing mucus, but no enzymes. External to the serosa is the **peritoneum**, which is a double-layered membrane surrounding the gut and lining the abdominal cavity. The **mesenteries**, which are extensions of the peritoneum, hold the gut canal in place by anchoring it to the abdominal wall. The surfaces of the cells of the peritoneum are moist, preventing frictional damage as the gut moves against other organs in the abdominal cavity.

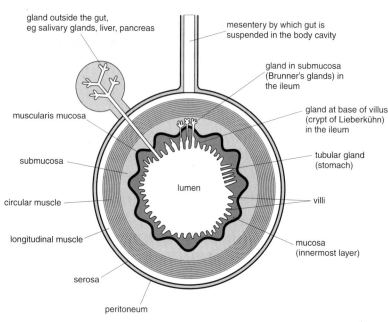

Figure 1.19 General structure of the gut wall

At certain regions along the gut are ducts from external glands, such as the salivary glands, the liver and the pancreas. Figure 1.19 gives a general plan of the gut wall with detail of specialised areas.

Mechanical digestion

In mechanical digestion, the large particles of ingested food may be sliced, crushed or otherwise broken up by the teeth. This reduces the food to smaller lumps which are easier to swallow and have a larger surface area, so that enzyme action during chemical digestion is more effective.

Mechanical digestion of ingested food is referred to as **mastication** and is achieved by means of teeth in the mouth, or buccal cavity. Teeth are present in the upper and lower jaws and in an adult human there are 32 permanent teeth. These consist, in each jaw, of: four **incisors**, two **canines**, four **premolars** and six **molars**, arranged as shown in Figure 1.20.

Crushing and grinding of the food not only breaks up larger lumps into smaller pieces, but also mixes them with saliva, which moistens and lubricates them. The tongue and cheek muscles also help to form the moistened food into a mass called a **bolus**, which is manipulated to the back of the buccal cavity before being swallowed.

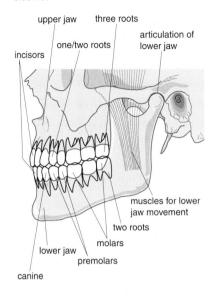

Side view

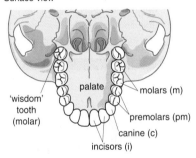

Surface view

Figure 1.20 Arrangement of adult human teeth

ADDITIONAL MATERIAL

The exposed part of each tooth, the **crown**, projects above the gum and is covered with **enamel**, which is very hard and resistant to decay. The **root** is embedded in the jawbone and is held in place by **periodontal fibres**, connected to the jawbone at one end and to the **cement** surrounding the outside of the root at the other. The **neck** is the part which is surrounded by the gum, but not embedded in the jawbone.

Most of the tooth consists of **dentine**, a hard, bone-like substance, which has many tiny channels extending through it. These channels, known as **canaliculi**, contain strands of cytoplasm from the dentine-producing cells in the pulp cavity, which is located in the central part of the tooth. In addition to the dentine-producing cells, the pulp cavity contains nerve endings and blood vessels (Figure 1.21).

The incisors have flattened crowns with sharp, chisel-like edges, which are used to bite off lumps of food. The canines have conical, pointed crowns and assist in the biting process. In carnivorous animals, the canines are much bigger and more pointed and are used to pierce and hold live prey. Both the incisors and the canines have single roots. The cheek teeth (the premolars and molars) have rounded projections, called **cusps**, on their crowns. These assist in crushing and grinding food during the process of chewing. Typically the cheek teeth each have more than one root, the molars on the upper jaw possessing three or four, so giving greater anchorage in the jawbone.

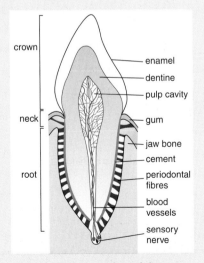

Figure 1.21 Section through human canine tooth

DEFINITION

Mastication is the term used to describe the crushing, slicing and breaking up of large masses of ingested food by the teeth. During this process large pieces of food are reduced to smaller lumps, so that they are easier to swallow.

17

EXCHANGES WITH THE ENVIRONMENT

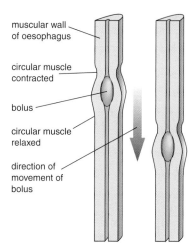

muscular wall of oesophagus

circular muscle contracted

bolus

circular muscle relaxed

direction of movement of bolus

Figure 1.22 Diagram to illustrate peristalsis

Movement of food along the alimentary canal

Associated with the muscularis externa are networks of nerves, one of which controls the contraction and relaxation of the muscles which bring about the movement of food along the alimentary canal. This movement, known as **peristalsis**, is initiated by the act of swallowing (Figure 1.22). The bolus stretches the wall and the circular muscle behind the bolus contracts, pushing the food forward. The circular muscle in the region surrounding the bolus is relaxed, increasing the diameter of the lumen and enabling the food to be pushed forward. Peristaltic movements occur all the way along the gut. In the small intestine, the alternate contraction and relaxation of the muscles brings about **segmentation movements** which help bring the contents of the lumen of the gut in contact with the epithelium of the wall where absorption occurs.

At certain points along the alimentary canal, the circular muscle is thicker and forms rings of muscle called **sphincters**, which control the passage of food from one region to another. The **cardiac sphincter**, found between the oesophagus and the stomach, controls the entry of food into the stomach. At the other end of the stomach, the **pyloric sphincter** relaxes to allow the passage of food from the stomach into the duodenum when it has reached the right consistency. There are additional sphincters at the junction of the ileum and the caecum, and at the anus.

Chemical digestion

Chemical digestion involves the action of **hydrolases** (hydrolytic enzymes) on the food constituents. The chemical breakdown of the complex organic molecules takes place progressively in stages until simple, smaller, soluble molecules are formed, which can then be absorbed. There are three major groups of digestive enzymes involved:

- **carbohydrases** hydrolyse the glycosidic bonds in carbohydrates
- **proteases** hydrolyse the peptide bonds in proteins and polypeptides
- **lipases** hydrolyse the ester bonds in triglycerides.

The processes which take place are the reverse of the condensation reactions which occur when carbohydrates, proteins and lipids are formed.

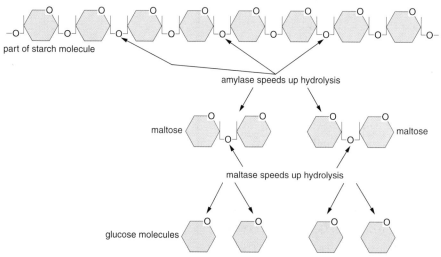

part of starch molecule

amylase speeds up hydrolysis

maltose

maltose

maltase speeds up hydrolysis

glucose molecules

Figure 1.23 Hydrolysis of starch to maltose and glucose, catalysed by amylase and maltase

Carbohydrases

Food does contain some monosaccharide sugars, such as glucose and fructose, but the bulk of carbohydrate in the human diet is in the form of polysaccharides and, therefore, requires digestion. Starch from plants and glycogen from animal sources are split into smaller units, usually disaccharides, by **amylases**. The disaccharides are then hydrolysed to monosaccharides which are small enough to be absorbed (Figure 1.23).

Carbohydrate digestion occurs in the mouth, duodenum and ileum under alkaline conditions. Carbohydrate digestion in humans is summarised in Table 1.3.

Table 1.3 *Carbohydrate digestion in humans*

Location in alimentary canal	Optimum pH	Enzymes involved	Products of digestion
mouth and buccal cavity	7.0	salivary amylase from salivary glands	dextrins (short chains of glucose residues), some maltose; food does not remain here long enough for much digestion
duodenum	7.0	pancreatic amylase from the pancreas	starch broken down to maltose
ileum (bound to membranes of microvilli on the epithelial mucosa)	8.5	maltase lactase sucrase from cells in small intestinal mucosa	maltose to glucose; lactose to glucose and galactose; sucrose to glucose and fructose

In the mouth, saliva is produced from three pairs of salivary glands (humans normally secrete about 1.5 dm³ each day). Saliva consists mostly of water, together with enzymes, mineral ions and mucus. The constituents of saliva and their roles are summarised in Table 1.4.

Table 1.4 *Constituents of saliva and their roles*

Constituent	Role
salivary amylase	initiates the digestion of cooked starch to maltose
lysozyme	catalyses the breakdown of cell walls of some pathogenic bacteria
mucus	helps to stick the food together to form a bolus
chloride ions	activate salivary amylase
other mineral ions (phosphates, hydrogencarbonate)	help to maintain the correct pH (about pH 7)

> **QUESTION**
>
> Explain why salivary amylase does not break down much starch.

Very little starch digestion takes place in the mouth, because the food does not remain there long enough. When a bolus of food is swallowed, it is transported down the oesophagus to the stomach, where the extremely acid conditions inhibit the action of salivary amylase. No carbohydrases are present in the gastric juice from the gastric glands in the mucosa of the stomach wall.

When the food reaches the duodenum, pancreatic juice containing amylase enters through the pancreatic duct. In addition to proteases and lipases, pancreatic juice also contains alkaline salts which help to neutralise the acid from the stomach. Bile from the liver is also added via the bile duct. It contains hydrogencarbonate ions which contribute to the creation of alkaline conditions in which the enzymes from the pancreas and on the surface of cells in the duodenum and ileum work most effectively. Pancreatic amylase hydrolyses any remaining starch to maltose.

The enzymes involved in the hydrolysis of disaccharides, such as maltose, lactose and sucrose, are located on the membranes of the microvilli of the epithelial mucosa. When hydrolysis is completed, the monosaccharides can then be absorbed.

ADDITIONAL MATERIAL

Proteases

There are two groups of proteases:
- endopeptidases speed up the hydrolysis of peptide bonds within the protein molecules
- exopeptidases act on terminal peptide bonds (those at the ends of the polypeptide chains).

Hydrolysis involving endopeptidases results in proteins being broken down into short polypeptide chains (Figure 1.24). Pepsin, trypsin and chymotrypsin are endopeptidases, each only capable of hydrolysing specific peptide bonds. For example, trypsin catalyses the hydrolysis of peptide bonds which involve the amino acids lysine or arginine. Pepsin, trypsin and chymotrypsin are secreted into the alimentary canal in their inactive forms: pepsin as pepsinogen, trypsin as trypsinogen and chymotrypsin as chymotrypsinogen. This ensures that these enzymes are activated only when there is food requiring digestion in the alimentary canal and prevents the enzymes damaging the cells in which they are produced. Pepsinogen is converted to pepsin by the action of hydrochloric acid in the stomach. Once some pepsin has been formed, it will bring about the conversion of more pepsinogen to pepsin. Trypsinogen is converted to trypsin by the action of the enzyme enterokinase, which is secreted in the ileum, and chymotrypsinogen is activated by trypsin.

It is also worth noting that some young mammals produce the enzyme rennin, secreted as pro-rennin, in the stomach. It is activated by hydrochloric acid and its function is to coagulate the soluble milk protein, caseinogen, to the insoluble calcium salt of casein, which is then hydrolysed by pepsin.

The action of exopeptidases results in the breakdown of the short polypeptide chains by the removal of amino acids. There are two kinds of exopeptidases:
- aminopeptidases hydrolyse peptide bonds at the amino end of a polypeptide chain
- carboxypeptidases hydrolyse peptide bonds at the carboxyl end of a polypeptide chain.

Protein digestion begins in the stomach, where the optimum pH for hydrolysis by pepsin is 1.5 to 2.0. Here the proteins and long polypeptide chains are broken down to shorter polypeptides. In the duodenum and ileum, where the pH is alkaline, trypsin and chymotrypsin from the pancreatic juice hydrolyse the proteins to shorter polypeptides. Carboxypeptidases are present in the pancreatic juice and their action results in the production of amino acids. Aminopeptidases are present on the microvilli of the epithelial mucosa.

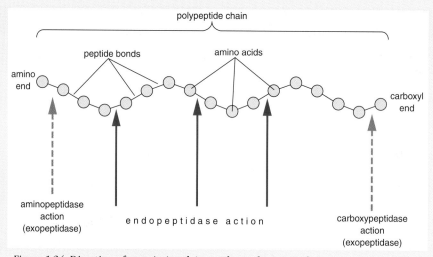

Figure 1.24 Digestion of protein involving endo- and exopeptidases

ADDITIONAL MATERIAL

Lipases

Most of the lipids in a human diet are triglycerides and, before they can be digested, they need to undergo emulsification into droplets. Bile salts from the liver, mostly sodium glycocholate and sodium taurocholate, are released into the duodenum via the bile duct from the gall bladder, where they have been stored. These salts lower the surface tension between the oil globules and water, bringing about emulsification and providing a larger surface area for the action of lipases (Figure 1.25).

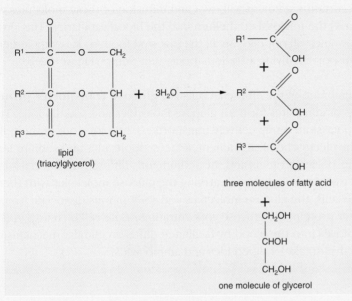

Figure 1.25 Hydrolysis of lipids catalysed by lipase

Lipases are secreted by the pancreas and released into the duodenum in the pancreatic juice. They catalyse the hydrolysis of triglycerides into monoglycerides, fatty acids and glycerol.

Absorption and histology of the ileum wall

Most absorption occurs in the small and large intestines, although water and alcohol can be absorbed from the stomach. The digested food is absorbed in the small intestine and water is mostly taken up from the large intestine.

The first 20 cm of the small intestine is known as the **duodenum** and it is here that the secretions from the liver and the pancreas are added. The rest of the small intestine is called the **ileum** and is about 5 m long. As we already know, the final stages of digestion of carbohydrates, lipids and proteins take place in the duodenum and ileum, whilst at the same time the process of absorption is occurring. There are several features of this region of the gut which contribute to the efficiency of absorption.

- It is long, providing a large surface area over which absorption can occur.
- There are large numbers of finger-like projections, the **villi** (Figure 1.26), in the mucosa, increasing the surface area for absorption.
- The villi possess smooth muscle fibres, which contract and relax, mixing up the contents of the ileum and bringing the columnar epithelial cells of the absorptive surface into greater contact with the digested food.

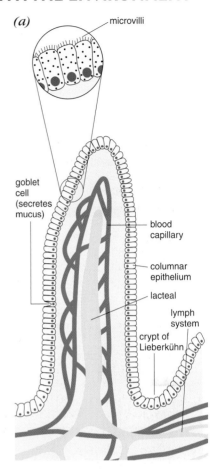

(a)

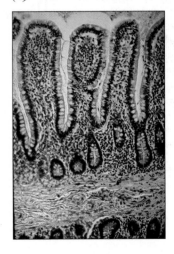

(b)

Figure 1.26 (a) Diagram to illustrate a vertical section through a villus; (b) photomicrograph of a section through the ileum showing villi

- The columnar epithelial cells possess **microvilli**, which further increase the surface area available for absorption.
- Each villus has an extensive capillary network so that the absorbed food is transported away quickly, maintaining concentration gradients.
- There is a lacteal in each villus into which absorbed fats pass.

Carbohydrates, in the form of monosaccharides, and amino acids are absorbed partly by diffusion and partly by active transport. Soon after a meal, there will be a higher concentration of monosaccharides and amino acids in the ileum, so a concentration gradient will exist and diffusion of these molecules will occur across the mucosal epithelium into the blood capillaries. This process is rather slow and will not take up all the digested food, so it is supplemented by active transport involving a sodium–potassium pump (Figure 1.27).

In the membrane of the epithelial cells is a **glucose transporter protein**, which has binding sites for both glucose molecules and sodium ions. The sodium–potassium pump actively transports sodium ions out of the cells against the electrochemical gradient. Glucose molecules and sodium ions bind to the transporter proteins and the sodium ions diffuse into the cells down their electrochemical gradient, carrying the glucose molecules with them. Inside the cells, the glucose molecules and sodium ions dissociate from the transporter protein, the glucose concentration of the cell increases and glucose moves into the blood by facilitated diffusion. Similar mechanisms exist for the active uptake of dipeptides and amino acids.

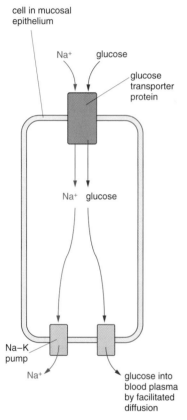

Figure 1.27 Diagram to illustrate glucose transporter protein and sodium–potassium pump in operation in a mucosal epithelial cell

ADDITIONAL MATERIAL

Digestion and absorption of fats

The absorption of the products of fat digestion, namely fatty acids and glycerol, involves a different mechanism. If the fatty acids are short, they can diffuse directly into the blood from the epithelial cells. The longer chain fatty acids, monoglycerides and glycerol diffuse into the epithelial cells of the mucosa, where they recombine to form fats. The fatty acids and monoglycerides have polar heads and non-polar tails, so they tend to clump together to form spherical structures, called **micelles**, in the lumen of the gut. Within a micelle, the non-polar tails point towards the middle and the polar heads are on the outside. Micelles can diffuse into the mucosal epithelial cells, but do so more slowly than single fatty acids. The molecules of fat diffuse into the lacteals, where they become coated with proteins present in the lymph to form droplets of lipoprotein called **chylomicrons**. These droplets are passed into the bloodstream from the lymph, where they are hydrolysed back into fatty acids and glycerol by an enzyme present in the blood plasma. They are then transported in the blood, from where they may be taken up and used as respiratory substrates by cells or stored as fat.

Mineral ions, vitamins and water are also absorbed from the contents of the duodenum and the ileum. Any remaining mineral ions and large amounts of water are absorbed from the food residues as they pass through the large intestine. Present also in this region are large numbers of bacteria which synthesise amino acids and vitamins, some of which are of use to humans and can be taken up and absorbed into the bloodstream.

Histology of the ileum wall

The structure of the ileum wall is similar to the general structure of the alimentary canal shown in Figure 1.19 on page 16. More detail, including the structure of a villus, is shown in Figure 1.26 on page 21. In the ileum region, the surface layer of the mucosa consists of columnar epithelial cells with large numbers of mucus-secreting goblet cells and the submucosa contains large amounts of lymphoid tissue.

The constant passage of food through the ileum damages the tips of the villi and cells are lost from the mucosa. These cells are replaced by cells from the bottom of the crypts of Lieberkühn situated at the base of the villi. These cells migrate to the surface and gradually move up the villus, eventually being shed from the tip. New cells are produced by mitotic divisions of the crypt cells. It has been estimated that between 50 and 200 g of the intestinal mucosa are renewed every day in an adult human. It takes from 5 to 7 days for a cell from the bottom of the crypt to move up to the tip of a villus.

2 Transport systems

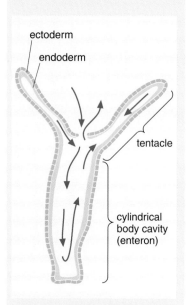

Figure 2.1 **Hydra** *showing two-layered body wall (the arrows indicate the circulation of water)*

The need for transport systems

In addition to the exchange of materials between an organism and its environment, which was discussed at the beginning of Chapter 1, there is the need for materials to be transported within the organism. Oxygen and nutrients need to be transported from their place of uptake to the respiring cells; carbon dioxide and other waste products must be removed. In green plants, water and mineral ions needed for photosynthesis are taken up by the roots and need to be transported to the leaves, where light and carbon dioxide are absorbed. The resulting organic compounds must be transported away from their site of synthesis to other regions for use in metabolic activities or for storage. The transport system in flowering plants consists of the vascular tissue, composed of xylem and phloem.

In mammals, absorption of the digested food takes place in the small intestine, from where it is transported to the liver and then to the respiring cells. Absorption of oxygen and removal of carbon dioxide occurs in the lungs, so a transport system is needed to deliver oxygen to the respiring cells and to remove carbon dioxide.

In many small organisms, efficient internal transport of all materials can be achieved through diffusion or active transport, because the distances involved are small and, where diffusion is concerned, the concentration gradients are favourable. In larger organisms, the distances between the different parts are too large and these processes are too slow.

The transport of respiratory gases in relation to the surface area to volume ratio has been discussed in Chapter 1. Increase in the bulk of an organism reduces the surface area available for the exchange of materials. Larger organisms possess internal transport systems to ensure the efficient delivery of materials needed for cell metabolism and the removal of the waste products.

In many animal groups, internal transport systems are present. These usually consist of a fluid, blood, which is pumped around the body by one or more muscular structures called hearts.

EXTENSION MATERIAL

Types of circulatory system

There are two types of closed circulatory system (Figure 2.2):

- **single circulations** – present in fish, in which blood is pumped from the heart to the gills, then to the rest of the body before being returned to the heart
- **double circulations** – typical of mammals, where blood is pumped from the heart to the lungs and back to the heart in a pulmonary circulation, followed by the systemic circulation in which the blood is pumped from the heart to the body organs and back to the heart.

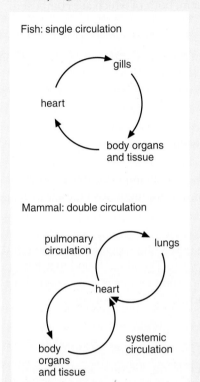

Figure 2.2 Fish and mammal circulations – to show differences between single and double circulatory systems

The concept of mass flow

Long-distance transport of fluids in living organisms takes place by **mass flow**. The fluid, together with any substances dissolved or suspended in it, moves in bulk, in response to a pressure gradient. This can be illustrated by considering the flow of liquid up a drinking straw. The person sucking the drinking straw lowers the pressure at the top end, which causes the liquid to move from a high pressure to a lower pressure area. In plants, movement of water and mineral ions up the xylem from the roots to the leaves is largely brought about by the evaporation of water from the leaves. This causes a lowering of the pressure at the top of the plant and does not involve metabolic energy.

Transport in flowering plants

In flowering plants, water, mineral ions and organic solutes are transported from their sites of uptake or synthesis to where they are used or, in the case of water, removed. This transport can occur over short distances from cell to cell, or over greater distances involving the vascular tissue, which is specialised for conducting water and solutes. Movement of substances from cell to cell may involve diffusion or active transport.

The structure of the vascular tissues

The vascular tissues are the xylem and phloem, both of which are composed of several distinct types of cells.

Water and mineral ions are transported from the roots to the aerial parts of the plant in the **xylem** tissue. This tissue may contain:

- vessels
- tracheids
- fibres
- unspecialised parenchyma cells, called xylem parenchyma.

The cell walls of tracheids, fibres and vessels have an impermeable substance called lignin deposited on them during their development, losing their living contents.

Both the tracheids and the vessels provide conducting tissue for the water and mineral ions, in addition to contributing support. The fibres found in xylem tissue are very similar to those found in supporting tissues in leaves and other non-woody structures. They have no conducting function, but contribute significantly to the strength of the tissue.

Tracheids (Figure 2.3) are elongated single cells with lignified walls and tapered ends. As they develop, they lose their living contents, so the lumen (cavity) of each cell is empty when mature. The thickening of lignin may take the form of rings, spirals, scalariform (consisting of interconnecting bars of lignin) or reticulate (more connections than scalariform), similar to those found in vessels. Typically, tracheids have pits, allowing the rapid transport of water from cell to cell.

Vessels (Figure 2.4) are found abundantly in the xylem of flowering plants and are long tubular structures with lignified walls. The vessels are formed by the joining of vessel segments, or vessel elements, end to end, the end walls of each vessel segment breaking down and leaving a perforation plate. The lignification can occur as rings, spirals, scalariform or reticulate, as described for tracheids. The rings and spirals occur more frequently in the first-formed xylem, thus allowing for the limited amount of stretching which might occur in young structures.

Organic solutes, which are the products of photosynthesis, are transported from their sites of synthesis in the leaves to other regions of the plants through the **phloem**. This tissue, like xylem, contains four types of cells:

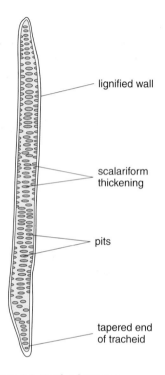

Figure 2.3 Tracheid structure

labels: lignified wall; scalariform thickening; pits; tapered end of tracheid

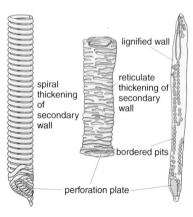

Figure 2.4 Vessel structure

labels: spiral thickening of secondary wall; lignified wall; reticulate thickening of secondary wall; bordered pits; perforation plate

- sieve tube elements
- companion cells
- phloem fibres
- phloem parenchyma (unspecialised parenchyma cells).

The sieve tube elements and companion cells (Figure 2.5) are involved in the long-distance transport of the organic solutes. The phloem parenchyma may serve as packing tissue, but some cells become modified to form **transfer cells**, which are responsible for the loading of the sieve tube cells and thus achieve the transport of the organic molecules over short distances.

Sieve tube elements are highly specialised living cells, linked end to end to form sieve tubes. The end walls of each of the sieve tube elements are perforated by **sieve plates**, so that long conducting tubes are formed. During development, each sieve tube element loses its nucleus and ribosomes, the vacuolar membrane breaks down and the remaining organelles, consisting of a few mitochondria and plastids, are distributed around the periphery of the cell. The cell surface membrane remains intact and large amounts of a special phloem protein, referred to as as **P-protein**, are formed. This protein may be organised into fibrils and it occupies most of the interior of the cell. It is not known precisely how the fibrils are arranged.

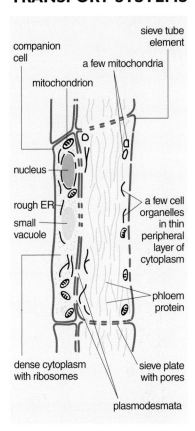

Figure 2.5 Sieve tube element and companion cell

EXTENSION MATERIAL

Sieve tubes in prepared slides

The contents of the sieve tubes are under high hydrostatic pressure, so that when specimens of tissue are observed, it is difficult to know whether the organelles have been disturbed by the sudden release of this pressure during preparation. In some preparations, the P-protein is seen to be blocking the pores in the sieve plates, whilst in others it does not. Another material, called **callose**, an insoluble polysaccharide, is often found deposited around the sieve plates. Again, it is difficult to know whether this material is found naturally in such situations or whether it is an artefact, formed during the procedures undergone in the preparation of the specimens. Callose is known to be formed when tissue is damaged.

Companion cells are closely associated with the sieve tube elements. Each companion cell comes from the same parent cell as its neighbouring sieve tube element. In contrast with the sieve tube elements, companion cells retain their nuclei, ribosomes and other organelles. The cytoplasm is dense and there are large numbers of mitochondria, indicating the potential for high levels of metabolic activity. There are numerous plasmodesmata connecting the cytoplasm of the companion cell with its neighbouring sieve tube element. The companion cells may provide replacement P-protein, enzymes and energy for activators of the sieve tube elements. Some companion cells appear to act as **transfer cells** and are involved in the movement of solutes into and out of the sieve tubes.

The general function of the transfer cells (Figure 2.6) appears to be the collection and transfer of organic solutes and inorganic ions into the sieve tubes. The inner

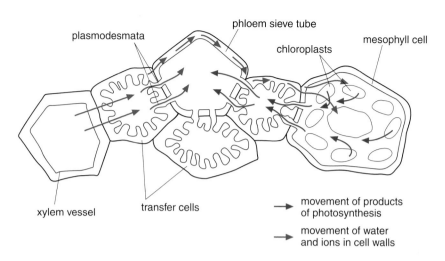

Figure 2.6 Transfer cells

walls of these cells have many folds and there are large numbers of mitochondria close to the folds. The folding increases the surface area available for the uptake of ions and solutes and the large numbers of mitochondria are able to supply the adenosine triphosphate (ATP) necessary for the active transport which is involved.

Movement of water
Water relations of plant cells

Plant cells, such as those in the epidermis and cortex regions of the roots, have living contents surrounded by a cell surface membrane and a thick, relatively inelastic cellulose cell wall. The cell wall gives special properties to these cells as it resists the osmotic uptake of water. A plant cell placed in distilled water will not swell up and burst, but will take up water until the pressure exerted by the cell wall prevents any further expansion. A plant cell in this condition is said to be **fully turgid**. Turgor is important for maintaining mechanical support in plants and, if plant cells lose water, the plant may wilt.

We use the term **water potential** to describe the force acting on water molecules in a solution, when separated from pure water by a membrane which only allows water to pass through it. Water potential is given the symbol ψ (Greek psi) and is measured in units of pressure, kilopascals (kPa). By definition, the water potential of pure water is zero. Adding a solute, such as sucrose, to pure water will decrease the water potential: it becomes negative. The more solute molecules present in a solution, the lower (or more negative) the water potential becomes. This change in water potential due to the presence of a solute is referred to as the **solute potential** and is given the symbol ψ_s.

Plant cells contain various solutes, such as sugars, which will exert a solute potential, so plant cells placed in distilled water will tend to take up water. This uptake is opposed by the inward pressure exerted by the cell wall. This pressure is known as the **pressure potential**, given the symbol ψ_p, and because it opposes the solute potential, it usually has a positive value. The overall water relationships of a plant cell can be summarised in the following equation:

$$\psi = \psi_s + \psi_p$$

water potential = solute potential + pressure potential

In plant tissues, water always tends to move, by osmosis, from a region of high water potential to a region of low water potential, down a water potential gradient.

In a fully turgid cell, the overall water potential of the cell is zero because the values of ψ_s and ψ_p are equal and opposite, so they cancel each other out. If a turgid cell is placed in a concentrated sugar solution, water will leave the cell because the solute potential of the sugar solution is much lower (more negative) than the water potential of the cell. As the cell loses water, the volume of the cell decreases and eventually the cell surface (plasma) membrane may lose contact with the cellulose cell wall. In this condition, the cell is said to be **plasmolysed**. The point at which the cell surface membrane is just about to lose contact with the cell wall is known as the point of incipient plasmolysis. At this point, the pressure potential is zero, so the water potential of the cell is equal to its solute potential. Plasmolysis can be induced experimentally by placing suitable cells, such as small pieces of epidermis from a rhubarb petiole, into concentrated sugar solutions, but it does not occur under natural conditions.

DEFINITION
Dicotyledons: A group of flowering plants in which the seeds have two cotyledons, or seed leaves, e.g. buttercup.
Monocotyledons: A group of flowering plants in which the seeds have one cotyledon, or seed leaf, e.g. grasses.

The structure of a dicotyledonous root

Before describing the process of water uptake, it is important to understand the structure of the root and the pathways involved in water transport. A transverse section of a young dicotyledonous root is shown in Figure 2.7.

The layer of cells around the outside of the root is the **epidermis**. Some epidermal cells develop projections called **root hairs**. The central part of the

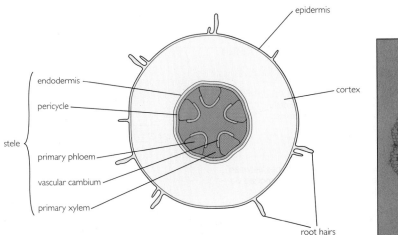
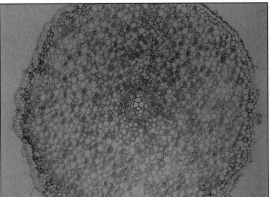

Figure 2.7(a) Transverse section of young Ranunculus *root; (b) photomicrograph of transverse section through a young* Ranunculus *root*

root contains the conducting tissues of the plant: the **xylem** and **phloem**. The xylem tissue provides a continuous system for the transport of water and dissolved mineral salts from the root, through the stem to the leaves. The function of the phloem is the transport of organic solutes, including sucrose which is formed in photosynthesis. The xylem and phloem are surrounded by a layer of cells known as the **pericycle**. The vascular tissues and their surrounding pericycle form a cylinder of conducting cells called the **stele**. Just around the outside of the stele is a layer of cells called the **endodermis**, which, as described later, has an important role in water movement in the plant.

Between the endodermis and the epidermis, there are several layers of relatively large, thin-walled cells forming the **cortex**. The cell walls of the cortical cells are highly permeable to water and dissolved solutes. There are also air spaces in the cortex which are important to allow oxygen to diffuse into the root for cell respiration.

Uptake of water

Water is taken up mainly by the younger parts of the roots (Figure 2.8), in the region of the root hairs. These are long projections (up to 15 mm) from the

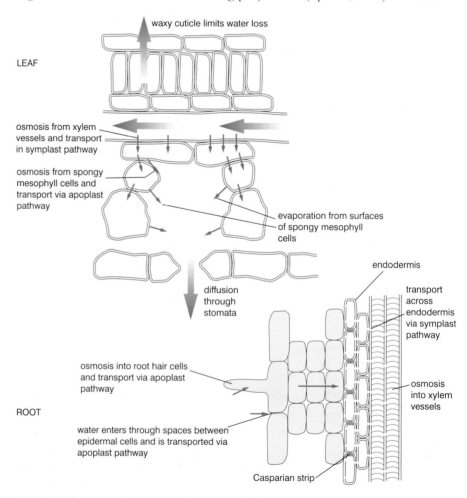

Figure 2.8 The transpiration stream showing uptake of water in a root and loss by evaporation from the leaf

epidermal cells which extend among soil particles and greatly increase the surface area for water uptake. The concentration of solutes, such as ions, is very low in the soil water of most soils, so the water potential of this water is close to zero. The solute potentials of plant cells are usually between about −500 and −3000 kPa, so there is a water potential gradient between root hair cells and soil water, and therefore water will be taken up by osmosis.

There are three ways in which water moves across the cortex of the root from the epidermis to the central tissues (Figure 2.9). These are:
* the **apoplast** pathway, in which water passes through the continuous system of adjacent cell walls
* the **symplast** pathway, in which water moves through the cytoplasm from cell to cell (the cytoplasm of adjacent cells in the cortex is in contact via the plasmodesmata, which are fine channels through the cell walls)
* the **vacuolar** pathway, in which water moves from vacuole to vacuole.

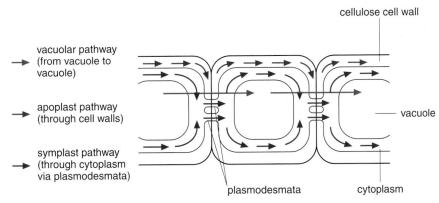

Figure 2.9 Apoplast, symplast and vacuolar pathways

However, when water reaches the endodermal cells its movement is stopped by a waterproof layer in the cell walls, called the **Casparian strip**. This is impregnated with suberin, a waxy compound, which is impermeable to water. Water is therefore prevented from passing around the endodermal cells through the cell walls, but instead it must pass through the cell surface membrane and cell contents. It is believed that, in this way, endodermal cells are able to regulate the movement of water and dissolved mineral salts from the soil into the xylem (see also page 36).

The transpiration stream
Water moves through a plant from the root hairs, where uptake occurs, through the root to the stem and leaves, where most of it is lost as water vapour in the process of **transpiration**. The continual evaporation of water from the aerial parts of the flowering plant into the atmosphere is due to a water potential gradient which exists between the soil and the atmosphere (Figure 2.10). The root hairs are in contact with the soil water, which has a high water potential, and the leaves and stems are exposed to the atmosphere, which has a much lower water potential. Consequently, there is a tendency for water to be drawn through and lost from the plant. This movement of water is a passive process. As water evaporates from the aerial parts of the plant, it

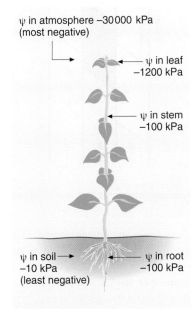

Figure 2.10 Water potential differences between soil and atmosphere

changes from a liquid to a vapour. This change of state requires heat energy, referred to as the **latent heat of vaporisation**.

Water vapour may be lost from three sites on the aerial parts of the plant:
- leaves (cuticle, stomata)
- flowers
- stems (cuticle, stomata; through spaces in the bark of woody stems).

Most of the water vapour loss occurs through the stomata, which are usually open during the day, allowing for the exchange of gases in photosynthesis. Evaporation through the cuticle, which accounts for about 10 per cent of the total water loss, will vary with its thickness. The amount of evaporation from woody stems is very small, but this is the main way in which water is lost from the stems of deciduous trees after leaf fall.

Most of the water lost by a plant is in the form of water vapour. Evaporation occurs from the cellulose cell walls of the mesophyll cells (mostly the spongy mesophyll) into the intercellular spaces, from where it diffuses out through the stomata, from a high water potential inside the leaf to a lower water potential outside. In dicotyledonous plants, there are usually larger numbers of stomata on the lower surfaces of leaves than on the upper, but in monocotyledonous plants like the grasses, with long, narrow leaves, the stomata are evenly distributed on both leaf surfaces.

Water reaches the mesophyll cells from the xylem of the vascular bundles in the leaf veins. These veins consist of only one or two vessels at the ends, with little lignification, so water can easily pass to the adjacent mesophyll via the apoplast, symplast or vacuolar pathways.

Removal of water from the xylem in the veins of the leaf creates a pulling force which draws water through the vascular tissue from the roots to the stem in continuous columns. The maintenance of these continuous columns is due to the structure of the xylem tissue and the properties of water.

In the xylem tissue, **tracheids** and **vessels** are the main conducting cells (see pages 26 and 27). Both tracheids and vessels have strong, rigid walls, which are able to withstand tension, and are small in diameter. There is an attraction between the water molecules and the walls of these cells (**adhesion**) and the water molecules stick to the walls. There are also strong forces between the water molecules (**cohesion**), which play a major role in maintaining the continuity of the water columns in the tissue.

Water columns under these forces are sometimes broken through wounding, entry of air, or pressure decreasing. Water will vaporise and the affected vessel will develop an air bubble, which blocks water movement. Due to the presence of pits (unlignified areas), which enable the sideways movement of water, the affected vessel may be bypassed (Figure 2.11).

Water movement in the xylem can be affected to some extent by **root pressure**. If the shoot of a plant is cut off close to the ground, liquid will exude from the

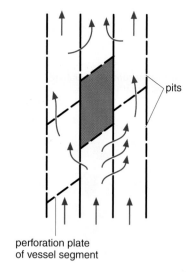

pits

perforation plate
of vessel segment

represents blocked
vessel segment

Figure 2.11 Vertical and sideways flow of water through the xylem

xylem tissue of the stump. This process occurs because ions are still being taken up actively by the roots and there is also osmotic uptake of water into the xylem. It can be demonstrated that a positive hydrostatic pressure of about 150 kPa may be generated by root pressure, so although it could not account for all the water movement in the xylem, it may have a contributory role. It has been suggested that air bubbles in the xylem of herbaceous plants may be removed by root pressure occurring when transpiration ceases at night.

Stomata and their role in transport

Stomata are pores in the epidermis of leaves, flowers and herbaceous stems through which exchange of gases occurs. They are found most frequently on leaves and may occur on both surfaces, but are more common on the lower (abaxial) surface. Surrounding each pore are two **guard cells** (Figure 2.12), which control the size of the opening by changes in their turgidity. The guard cells are usually kidney-shaped, but in grasses they are dumb-bell-shaped. The adjacent epidermal cells are often arranged in a characteristic pattern and are referred to as **subsidiary cells**. The guard cells are different from the rest of the epidermal cells in that they contain chloroplasts, they are not linked to adjacent cells by plasmodesmata and their walls are unevenly thickened. The part of the cellulose cell wall which borders the pore is thicker, and also less elastic, than the opposite wall. As water is taken into the guard cells, increasing their turgidity, they become more curved and the pore between them opens wider. When the guard cells lose turgidity, they become less curved and the pore closes.

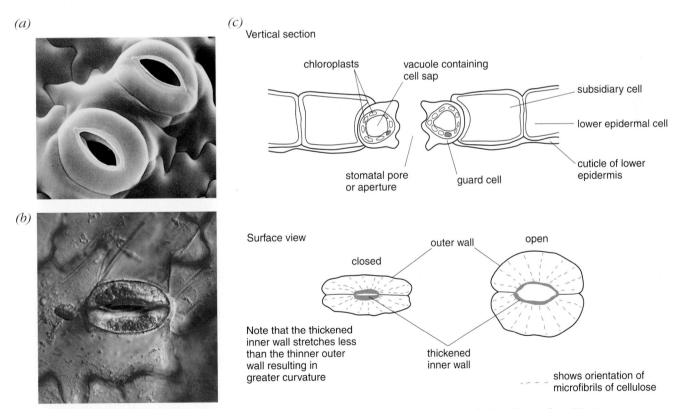

*Figure 2.12 Stomatal structure – (a) scanning electronmicrograph of open stomata on the leaf of a tobacco plant (*Nicotiana tabacum*); (b) light micrograph of closed stomata on the leaf of a broad bean (*Vicia faba*); (c) vertical section and surface view.*

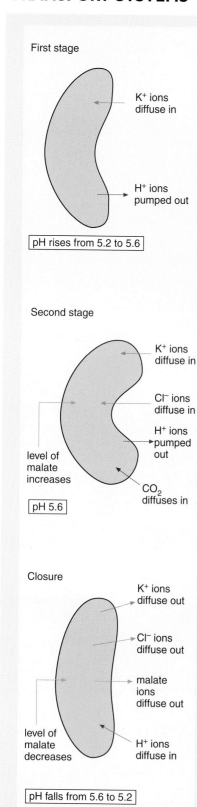

First stage

K⁺ ions diffuse in

H⁺ ions pumped out

pH rises from 5.2 to 5.6

Second stage

K⁺ ions diffuse in

Cl⁻ ions diffuse in

H⁺ ions pumped out

level of malate increases

pH 5.6

CO₂ diffuses in

Closure

K⁺ ions diffuse out

Cl⁻ ions diffuse out

malate ions diffuse out

level of malate decreases

H⁺ ions diffuse in

pH falls from 5.6 to 5.2

Figure 2.13 The chemiosmotic mechanism of stomatal opening and closing

The chemiosmotic mechanism of stomatal opening and closing

At the same time as the opening of a stoma, there is a considerable increase in the concentration of solutes in the guard cells, causing the water potential to become more negative (lower) (Figure 2.13). This causes water to move into the cells and the **turgor pressure** increases. Investigations of the concentrations of different ions in the guard cells of open and closed stomata have shown that, as the stoma opens, there is a steep rise in the concentration of potassium ions and chloride ions. The first stage in this process is thought to involve the removal of hydrogen ions from the guard cells due to the action of a **proton pump**. An electrochemical gradient builds up across the guard cell membrane and so the potassium ions diffuse in passively. Due to the active removal of hydrogen ions, the pH inside the guard cell increases, i.e. it becomes more alkaline whilst that outside decreases, becoming more acid.

A number of factors influence the opening and closing of the stomata. These include the light conditions, the supply of water to the plant and the supply of respiratory substrates. In plants that are well supplied with water, the stomata open at dawn and close at dusk and this pattern appears to be controlled by light and the amount of carbon dioxide in the intercellular spaces of the leaves. Photosynthesis is initiated by light and, as soon as the process begins, there will be a decrease in the amount of carbon dioxide in the leaf. It has been observed that low concentrations of carbon dioxide in the leaves promote stomatal opening and high concentrations bring about stomatal closure. Although it has been possible to show that the guard cells are sensitive to different levels of carbon dioxide, the precise mechanism of how this might cause stomatal opening is not known.

It is quite common for stomata to close around midday in leafy trees. When air temperatures are high and the humidity is low, transpiration is high and water loss may exceed water uptake. This closing of stomata could avoid air locks in the xylem vessels impeding the transpiration stream. It appears to be controlled by external factors, but if wilting is imminent, the stomata close rapidly. Under these circumstances, the decrease in turgor of the leaf cells triggers the synthesis of a plant growth inhibitor, **abscisic acid** (ABA), in the chloroplasts. If the level of ABA is sufficiently high, it affects the cell surface membrane of the guard cells, preventing the proton pump from operating and bringing about closure of the stomata. As soon as more water is available to the plant, the ABA is broken down.

ADDITIONAL MATERIAL

To control the intracellular pH, there is a second stage to the process, which involves the inward diffusion of chloride ions, either as a result of the change in pH or linked with hydrogen ion uptake. In addition, there is an increase in the amount of malate in the guard cell. It is thought that the increase in pH activates the enzyme phosphoenolpyruvate carboxylase (PEP carboxylase), which catalyses the fixation of carbon dioxide to produce oxaloacetic acid. The oxaloacetic acid can be reduced to malic acid, which dissociates, providing malate ions to balance the potassium ions and hydrogen ions for the proton pump. Starch stored in the guard cells provides the pyruvate, which is first converted to phosphoenolpyruvate before carbon dioxide fixation occurs.

Factors affecting the rate of transpiration

The rate of transpiration can be affected by environmental factors and also by a number of structural or internal features of the plants. The environmental factors include:

- light
- temperature
- humidity
- air movements.

Light affects the rate of transpiration because the size of the stomatal aperture is controlled by light. As stomata open in the morning, the rate of transpiration increases, decreasing at dusk when the stomata close. Some transpiration may occur through the cuticle, referred to as **cuticular transpiration**, when stomata are closed.

The rate of transpiration increases with an increase in **temperature**, because higher temperatures cause water to evaporate more rapidly from the cell walls of the mesophyll tissue. This increases the concentration of water vapour molecules in the air spaces in the leaf. Warmer temperatures also lower the humidity of the air outside the leaf, which increases the difference in water potential between the leaf and the atmosphere, so water will diffuse out more rapidly.

The water vapour pressure, or **humidity**, of the atmosphere has an effect on the rate of transpiration. If the humidity is low, the air is relatively dry and there is a steeper diffusion gradient between the external atmosphere and the atmosphere inside the leaf, so the rate of transpiration is higher. If the humidity is high, the air is more saturated with water vapour molecules and the opposite applies.

In still air, 'shells' of air saturated with water vapour molecules are built up around the leaves, which reduces the rate of transpiration because the diffusion gradients are less steep. Any air movement can disturb these shells, moving the water vapour molecules away from the surface of the leaves and thus creating steeper diffusion gradients.

Internal factors which affect the transpiration rate include:

- the surface area of the leaf – the greater the surface area, the higher the rate of transpiration
- the thickness of the cuticle – a thick cuticle reduces the rate of cuticular transpiration
- stomatal density – the greater the number of stomata per unit area of leaf, the greater the rate of transpiration.

Many plants living in dry conditions show adaptations which have the effect of reducing the rate of transpiration and thus conserving water. Such plants are referred to as **xerophytes** and their adaptations as **xeromorphic**. The adaptations shown by some xerophytic plants are described in Chapter 3.

QUESTION

What combination of environmental factors causes the highest rate of transpiration?

QUESTION

What additional features are typical of the leaves of xerophytes?

Movement of nutrients
Uptake of mineral ions, their transport and circulation

Mineral ion uptake in flowering plants takes place at the roots, the ions being absorbed along with water from the soil solution. Small amounts of ions can also be absorbed by the leaves from rainwater. Most of the uptake occurs in the root hair region of young roots (Figure 2.14), where there is a large surface area available for rapid uptake. If the concentration of ions outside the roots is higher than their concentration within the roots, ions may diffuse in passively via the apoplast pathway. This situation occurs in the case of calcium ions, which are nearly always in higher concentrations in the soil solution than in the root cells. The uptake of most ions, however, is an active process, against a concentration gradient. This is because the majority of ions needed are present in higher concentrations in the root cells than in the soil solution.

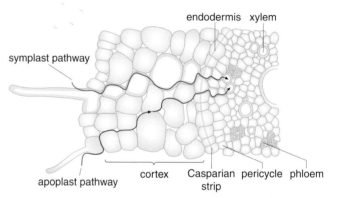

Figure 2.14 Symplast and apoplast pathways of ion absorption in the root hair region. The symplast pathway involves transport through the cytosol (stippled) of each cell all the way to the non-living xylem. The apoplast pathway involves movement through the cell wall network as far as the Casparian strip, then movement through the endodermis. Casparian strip of endodermis is shown only as it would appear in end walls.

Active transport requires a supply of energy to drive proton pumps in the cell surface membrane. Hydrogen ions are pumped out of the cells across the cell surface membrane, creating electrochemical gradients for selective ion uptake by specific transport proteins or carriers. Once inside the root cells, the ions are transported across the cortex to the endodermis via the symplast pathway.

Many of the cells of the endodermis have bands of impermeable suberin on their cell walls. These bands are referred to as **Casparian strips** (Figure 2.15) and prevent the passage of water and ions into the xylem via the apoplast pathway. Water and mineral ions must pass through the cell surface membrane and cytoplasm of these cells before entering the xylem. This has prompted suggestions that the endodermis can regulate the uptake of ions by the plant. Observations of transverse sections of roots indicate the presence of certain endodermal cells, referred to as 'passage cells', that lack Casparian strips and would allow water and mineral ions to move freely into the xylem.

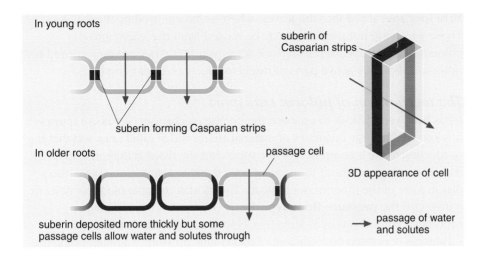

In young roots

suberin of Casparian strips

suberin forming Casparian strips

3D appearance of cell

In older roots

passage cell

suberin deposited more thickly but some passage cells allow water and solutes through

→ passage of water and solutes

Figure 2.15 Casparian strips in endodermal cells and passage cells

ADDITIONAL MATERIAL

Evidence for translocation in the phloem

Early investigations of phloem transport involved the removal of a ring of bark from the trunk of a tree to see what would happen. Such an experiment was carried out by Marcello Malpighi in 1679. He removed the bark and underlying soft tissues of the stem, leaving the xylem intact. After a few weeks, the tissue on the upper side of the ring was swollen while that below was not. Malpighi found that the swelling only occurred if there were green leaves present on the tree and if the tree was kept in the light, so he deduced that substances were being made in the light and transported down the stem. At the time, this was quite a novel idea, and was some indication that a special tissue was involved, independent from that in which water transport occurred.

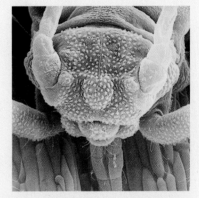

In the 1950s, more sophisticated techniques were developed involving the use of radioactive tracers. It was possible to supply a leaf with radioactively-labelled carbon dioxide ($^{14}CO_2$) in the light. After a period of time, stem tissue from the plant could be taken, frozen, dehydrated and cut into thin sections. If the sections were placed on photographic film, the position of any radioactively-labelled substances would show up when the film was developed. In this way, not only the tissue but also the cells involved in the transport of the organic substances could be identified.

Figure 2.16 Aphid feeding on phloem

Analysis of phloem exudates enables the contents of the phloem sap to be identified. It is possible to collect sap directly from some monocotyledonous species, such as palms, when they are cut, due to the high hydrostatic pressure forcing the sap out, but the use of aphids eliminates the possibility of contamination. Aphids are allowed to feed on a plant by inserting their stylets into the phloem tissue (Figure 2.16). Once penetration of the tissue has occurred, the aphids are anaesthetised to prevent them withdrawing their stylets and then their bodies are cut off, leaving the stylets still inserted into the phloem. Each stylet acts like a tiny pipette and sap can be collected over a number of days. This technique has its drawbacks as it is difficult to carry out and the amounts of sap collected are small. However, it can be combined with the use of radioactive tracers to provide useful information on the rate of translocation. If radioactively-labelled material is introduced, it is possible to measure the time taken for this material to be transported over a measured distance.

Using some of the techniques described above, it has been shown that phloem sap is an alkaline solution, consisting mostly of sucrose, together with organic nitrogen compounds and potassium ions. The concentration of sugars in the sap varies from 15 to 30 per cent, compared with 0.5 per cent in the leaf cells, indicating that the loading of the sieve tube elements is an active process, requiring energy. Any process or compound, such as a metabolic poison, which slows or inhibits respiration, will slow down or stop translocation in the phloem.

Most ions are carried into the leaves, where some are used by the leaf cells. Those which are not used here may be moved from the xylem into the phloem, through special transfer cells (see page 27). They are then carried to other metabolically active parts of the plant, such as developing buds.

The mechanism of phloem transport

It has been very difficult to establish the precise mechanism of phloem transport. It is known that large quantities of material are moved at rapid rates and that the conducting tissue is composed of fine tubes, but the tissue is delicate and easily damaged. It is also difficult to see what roles the P-protein and the sieve plates play in most of the theories put forward. The most acceptable model of phloem transport is the **pressure flow hypothesis** (Figure 2.17), based on an hypothesis put forward by E. Münch in 1930. He suggested that the movement of substances depended on a gradient of hydrostatic pressure. The fluid inside the sieve tubes in the leaves, at the **source** end of the system, has a higher hydrostatic pressure than in the sieve tubes in places such as the roots, developing buds or seeds, known as the **sink** end of the system.

In the model (Figure 2.17), A represents the source and B the sink. The solute concentration at A is higher than at B and so water enters A by osmosis. This causes a high hydrostatic pressure to build up, forcing water out of B. Mass flow of the contents of A along to B results, due to the hydrostatic pressure gradient.

If this model is applied to the situation in living plants, there would be high concentrations of sugars at A due to photosynthesis in the leaves. Water enters the leaf cells by osmosis, increasing their turgor pressure. It is known that sugars will be used for respiration and synthesis in other regions of the plant, represented by B, so solutes and water will move out, resulting in a lower hydrostatic pressure at B. Transfer cells actively transport sucrose into the phloem. This causes water to enter, producing a higher hydrostatic pressure at the source. At the sink, sucrose diffuses out to the respiring cells as the concentration is low in these cells. Water moves out by osmosis, lowering the hydrostatic pressure. Mass flow occurs along the phloem from the source to the sink.

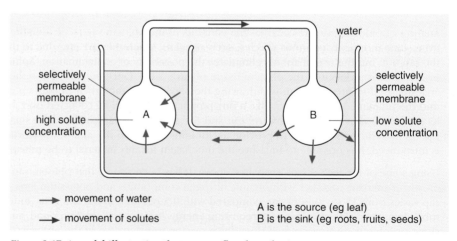

Figure 2.17 A model illustrating the pressure flow hypothesis

The pressure flow hypothesis predicts that a pressure gradient exists along sieve tubes and this has been demonstrated. It is not known whether the gradient would account for the rapid transport of materials in phloem tissue. The hypothesis also suggests that the water and solutes move together at the same rate in the same direction. In some of the experiments carried out with radioactively-labelled sugars and water, the sugars moved at a more rapid rate. It was suggested that, as the sieve tubes were permeable to water, some moved out into the surrounding tissues.

Münch's original hypothesis was a purely physical one and did not include any reference to living tissues or the possibility of active transport. It would seem necessary to modify the original hypothesis to account for the active transport of solutes into the sieve tubes in the leaf and the unloading of solutes at the sinks, but mass flow seems to account for the movement once the solutes have reached the phloem tissue.

Transport in humans and other mammals

The circulatory system

The function of the circulatory system is the transport of nutrients and other substances, including gases, hormones and excretory products, to and from various parts of the body. The blood must therefore be kept in a state of continuous circulation, the energy for which is provided by the heart. The force of contraction of the heart propels blood to the tissues through thick-walled **arteries** and back to the heart through the thinner-walled **veins**. In the tissues, blood passes through a network of **capillaries** in which exchange of materials occurs between blood and the tissue fluid.

Mammals have a double circulatory system, which consists of the **systemic circulation**, in which blood is pumped from the left side of the heart to the various tissues and organs of the body, and the **pulmonary circulation**, in which blood is pumped from the right side of the heart to the pulmonary capillaries of the lungs.

Structure and function of the heart

The human heart weighs about 300 g and is situated in the middle region of the thorax. The lower end of the ventricles, known as the apex, points towards the left of the thorax. The structure of the heart and major blood vessels is shown in Figures 2.18 and 2.19.

The wall of the heart contains a thick, muscular layer, called the **myocardium**. This consists of **cardiac muscle**, a special type of muscle tissue.

> ### QUESTION
> Animals such as fish have a single circulatory system, in which blood is pumped from the heart to the gills and then directly round the body. What are the advantages of a double circulatory system? (See page 25 for details.)

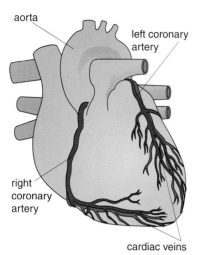

Figure 2.19 The coronary circulation.
Cardiac muscle cells receive their
blood supply by means of two main
arteries, the left and right **coronary
arteries**. These arise directly from the
aorta, immediately above the aortic
semilunar valve. Blood passes
through capillaries within the cardiac
muscle, then returns to the right
atrium via the **cardiac veins**

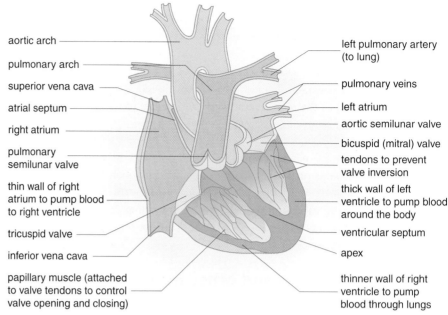

Figure 2.18 Internal structure of the heart

DEFINITION

Striated muscle is the type
of muscle tissue associated
with the skeleton and
movement. Unlike cardiac
muscle, striated muscle is
under conscious control so
that contraction of striated
muscle occurs voluntarily. As
a result, striated muscle is
sometimes referred to as
voluntary muscle. The biceps
and triceps muscles in the
upper arm, for example,
consist of striated muscle.

DEFINITION

Cardiac output is the
volume of blood pumped out
of either the left or right
ventricle per minute.

ADDITIONAL MATERIAL

Cardiac muscle consists of many branching
cells, each of which may contain one or two
nuclei. Cardiac muscle cells are joined by
structures known as **intercalated discs**,
specialised junctions between cells which
both transmit the force of contraction and
allow the rapid spread of electrical excitation
throughout the myocardium. In histological
preparations, intercalated discs appear as
dark, transverse lines, as seen in Figure 2.20.
Notice too that cardiac muscle has cross-
striations, similar to those of striated muscle.

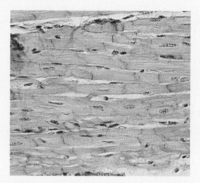

Figure 2.20 Histology of cardiac
muscle, as seen using a light
microscope

Cardiac muscle is said to be **myogenic**. This means that, unlike striated
muscle, cardiac muscle is self-exciting. Cardiac muscle cells show a continuous,
rhythm of electrical excitation and contraction on their own, although this can
be changed by nervous or hormonal influences.

Cardiac muscle has a very dense capillary network which receives blood by
means of the left and right **coronary arteries**. These are the first branches
from the aorta and, at rest, receive about 5 per cent of the total **cardiac
output**. After passing through the capillaries, blood returns mainly via a series
of cardiac veins which drain into the right atrium.

The interior of the heart is divided into four chambers: two upper **atria** and
two lower **ventricles**. The atria receive blood from veins which return it to the

heart. The atria contract and push blood into the ventricles, which then contract with considerable force and pump blood into the arteries. The myocardium of the ventricles is thicker than that of the atria and the myocardium of the left ventricle is much thicker than that of the right ventricle. The left ventricle pumps blood into the aorta and around the entire body at a higher pressure than the right ventricle, which pumps blood into the pulmonary arteries and through the pulmonary capillaries.

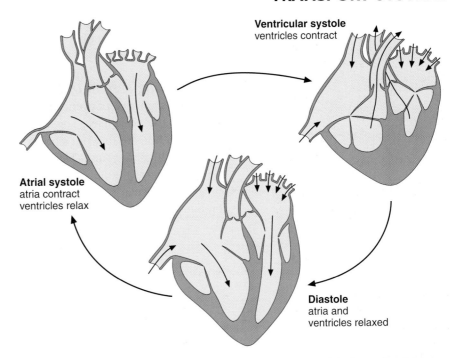

Figure 2.21 Chambers and valves of the heart showing the action of the heart chambers during atrial systole and ventricular systole

The rhythmic sequence of events which occurs each time the heart beats is known as the **cardiac cycle** (Figure 2.21). At rest, the heart beats about 72 times per minute, so each cycle lasts about 0.83 seconds. The cardiac cycle consists of:

- **atrial systole** – contraction of the atria
- **ventricular systole** – contraction of the ventricles
- **complete cardiac diastole** – relaxation of the atria and ventricles.

The heart valves

The heart valves ensure that blood flows in one direction only and are essential for the normal function of the heart. The **atrioventricular (AV) valves** are situated between the two atria and the ventricles. The right atrioventricular valve has three flaps and so is also known as the **tricuspid valve**. The left atrioventricular valve is similar in structure but has two, rather than three flaps, and is therefore called the **bicuspid**, or **mitral valve**. These valves prevent blood from flowing back into the atria when the ventricles contract and ensure that blood moves into the aorta and pulmonary arteries. The free edges of these valves are attached to papillary muscles, small projections of the inner walls of the ventricles, by chordae tendinae, which prevent the valves from opening upwards. The opening of the pulmonary trunk (which forms the left and right pulmonary arteries) is guarded by the **pulmonary semilunar valve** and the opening of the aorta is guarded by the **aortic semilunar valve**. When the semilunar valves close during diastole, blood is prevented from flowing back into the ventricles from the aorta and pulmonary arteries.

It is the closure of these valves which is responsible for the heart sounds. During the cardiac cycle, the atrioventricular valves close simultaneously at the start of ventricular systole. This produces the first heart sound, followed shortly afterwards by the second sound, produced by the closure of the semilunar valves indicating the start of diastole. The first and second heart sounds are frequently described as 'lubb' and 'dup' respectively and may be heard using a stethoscope.

The conducting system of the heart

The rhythmic sequence of events during the cardiac cycle is coordinated by tissues within the heart itself. These tissues consist of modified cardiac muscle cells which, in humans, are not easy to distinguish from other cardiac muscle cells. The conducting system (Figure 2.22) comprises:

- the sinoatrial (SA) node
- the atrioventricular (AV) node
- the bundle of His and its branches.

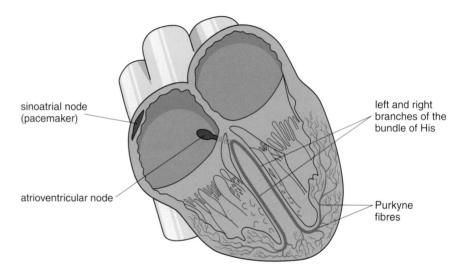

Figure 2.22 The conducting system of the human heart

The sinoatrial node (SA node) consists of a small group of specialised cells situated in the wall of the right atrium, near the opening of the superior vena cava (see Figure 2.18 on page 40). The SA node is often referred to as the **pacemaker**, because it initiates the heart beat by sending out a wave of electrical excitation that spreads over the right and left atria. This stimulates the cardiac muscle of both atria to contract.

The atrioventricular node (AV node) is situated in the wall of the right atrium, along the lower part of the wall that separates the right and left atria. The wave of excitation reaches the AV node and passes relatively slowly through it. This delays the spread of the excitation for a fraction of a second before it spreads to the ventricles, so that the atria have contracted fully before the ventricles start to contract. The excitation is conducted from the AV node, via the bundle of His and branches, to the cardiac muscle of the ventricles, which then contract from the apex upwards. The bundle of His and its branches consists of Purkyne (or Purkinje) tissue – modified cardiac muscle tissue. This arrangement of the conducting tissues ensures that there is a delay between contraction of the atria and contraction of the ventricles and that the electrical excitation reaches most of the cells in the ventricles at the same time, which ensures a single, coordinated contraction.

The electrical impulses which accompany contraction of the heart are conducted through body fluids and can be recorded by placing electrodes on the surface of the skin, either on the chest wall, or on the wrists and ankles. The pattern of electrical activity can then be displayed on an oscilloscope screen, or printed out onto paper. This recording is known as an **electrocardiogram**, or **ECG**, and shows changes in voltage against time. A normal ECG shows five waves, which are conventionally referred to as the P wave, the QRS complex and the T wave (Figure 2.23).

The P wave is caused by electrical excitation of the atria. The QRS complex indicates excitation of the ventricles and the T wave is due to recovery (repolarisation) of the ventricles. The ECG has considerable clinical importance as it indicates abnormalities in the pattern of excitation of the heart. Part of an ECG from a healthy patient is shown in Figure 2.23.

> **QUESTION**
>
> Identify the P, Q, R, S and T waves in Figure 2.23. How many complete cardiac cycles are shown?

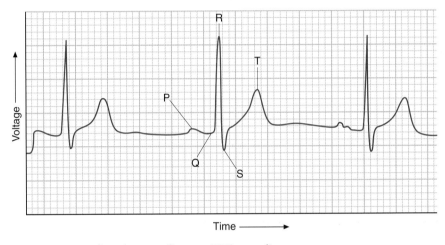

Figure 2.23 Part of an electrocardiogram (ECG) recording

Artificial pacemakers

On occasions a heart attack, resulting in the death of some cardiac muscle cells, causes damage to the conducting system of the heart. As a result, the atria may continue to beat normally but the ventricles, which no longer receive impulses via the bundles of His, start to contract, but much more slowly than is required to maintain adequate cardiac output. Patients may become very short of breath, or suffer from dizziness or blackouts. This condition can be treated by the insertion of an **artificial pacemaker**, a device which delivers an electrical stimulus to the heart muscle. There are several different types of artificial pacemakers, including permanent pacemakers, which are inserted into the patient's chest and connected to the apex of the heart via a pacing wire. The pacemaker is powered by lithium batteries and generates an electrical stimulus which ensures that the ventricles contract at a steady rate of 60 to 70 beats per minute. Some advanced types of pacemakers, known as rate-responsive pacemakers, can alter the rate of stimulation to match physiological demand. For example, the rate will increase during physical exercise.

Pressure changes in the heart

Figure 2.24 shows the pressure changes in the heart during a cardiac cycle, starting with the beginning of atrial systole at time 0. Notice that, although the pressure in the ventricles drops to zero during diastole, the pressure within the aorta and pulmonary artery remains relatively high. This pressure is maintained by the closure of the semilunar valves and by the elastic recoil of the arterial walls.

QUESTION

Study Figure 2.24 carefully. Compare the changes in pressure in the left ventricle with those in the right ventricle. Suggest reasons for the differences.
Explain why the pressure in the aorta remains relatively high when the pressure in the left ventricle falls to around zero.

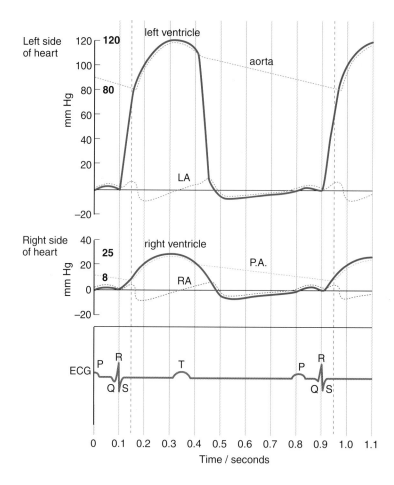

Figure 2.24 The sequence of events in the cardiac cycle. PA = blood pressure in the pulmonary artery; RA = pressure in the right atrium; LA = pressure in the left atrium. Blood pressure is usually expressed in mm Hg, but may be converted to kilopascals by dividing by 7.5.

Arteries, capillaries and veins

Arteries are blood vessels which transport blood away from the heart. Although arteries vary in diameter, they all have a similar structure, consisting of three layers of tissue:

- an inner tunica intima, consisting of flattened epithelial cells (endothelium) and their supporting connective tissue
- a middle tunica media, containing smooth muscle cells and elastic fibres
- an outer tunica adventitia, consisting of connective tissue.

Arterioles are defined as vessels of the arterial system with a diameter of less than 0.3 mm. Their tunica media consists almost entirely of smooth muscle cells. Arterioles divide repeatedly and become progressively smaller in diameter, eventually leading into **capillaries**.

The wall of a capillary consists of a single layer of flattened epithelial cells, which allow efficient exchange of materials between blood and the surrounding tissue fluid. The diameter of capillaries varies, but is typically about 7 μm, approximately the same as that of a red blood cell.

Capillary networks drain into venules, then **veins**, which return blood to the heart. The wall of a vein has the same three layers as the wall of an artery, but the wall is very much thinner in relation to the diameter of the lumen. Some veins contain valves: delicate projections of the tunica intima. These valves are present mainly in the veins of the limbs and prevent backflow of blood. The structures of an artery, capillary and vein are shown in Figure 2.25.

Blood and body fluids

Blood is a tissue consisting of a variety of cells suspended in a fluid known as **plasma**. Blood functions mainly as a means of transport throughout the body for respiratory gases (oxygen and carbon dioxide), nutrients, hormones, metabolic waste products and cells. Plasma consists of 90 per cent water and 10 per cent solutes, including proteins, nutrients, excretory products, dissolved gases, hormones, enzymes and other substances. Plasma proteins are of three types: albumins, globulins and fibrinogen. Albumins and fibrinogen are important in the blood-clotting mechanism; globulins are essential parts of the immune system.

There are three varieties of cells present in blood:
- erythrocytes, or red blood cells
- leucocytes, or white blood cells
- thrombocytes, or platelets.

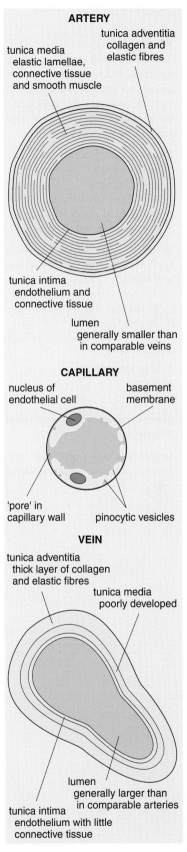

Figure 2.25 Structure of blood vessels (not to same scale)

ADDITIONAL MATERIAL

Blood pressure

The term **blood pressure** usually refers to the pressure within the aorta and main arteries. The pressure is not constant, but varies between a minimum (or diastolic) value and a maximum (or systolic) value. Blood pressure is usually measured in mm Hg, although this is not an SI unit. Typical blood pressures for an adult at rest are about 120 mm Hg (systolic) and about 80 mm Hg (diastolic), usually written as 120/80 mm Hg. However, it must be noted that these figures will vary according to the time of day (there is a circadian rhythm of changes in blood pressure), posture, sex and age of the person.

Blood pressure depends on a number of factors including:
• cardiac output
• blood volume
• peripheral resistance (resistance to blood flow mainly due to the diameter of arterioles)
• elasticity of artery walls
• volume of blood returning to the heart.

As an example, cardiac output depends on the volume of blood pumped out of the ventricles with each beat and the heart rate. Any factor which makes the heart beat faster, or makes it beat more strongly and increases the stroke volume, will increase cardiac output and will therefore tend to increase blood pressure. High blood pressure, or **hypertension**, occurs when the diastolic pressure exceeds 95 mm Hg and the systolic pressure exceeds 160 mm Hg (World Health Organisation classification). Many risk factors have been identified in the development of hypertension and include:
• genetic factors
• gender – men experience higher rates of hypertension at an earlier age than women
• age
• high stress levels
• obesity
• smoking.

There are numerous complications of untreated hypertension, including heart failure, kidney failure, atherosclerosis and stroke.

Hypotension is a relatively rare condition in which the systolic blood pressure is typically below 105 mm Hg (14 kPa) and the diastolic pressure less than 60 mm Hg (8 kPa).

All blood cells develop from cells known as stem cells, which are present in bone marrow. Bone marrow is found inside bones, particularly in long bones and between the layers of bone in flat bones such as the sternum, ribs and skull. Stem cells divide by mitosis and give rise to separate cell types, each of which divides and develops to form one of the main blood cell types.

White blood cells are subdivided into two main groups: **granulocytes** and **agranulocytes**. Granulocytes have granules in their cytoplasm and a multilobed nucleus. There are three types of granulocytes: **neutrophils**, **eosinophils** and **basophils**. Agranulocytes, which are either **monocytes** or **lymphocytes**, have nongranular cytoplasm and their nuclei are not lobed, although the nucleus of a monocyte may be strongly indented. The structures and functions of blood cells are summarised in Table 2.1 on page 48.

Transport of oxygen and carbon dioxide

Oxygen combines with haemoglobin, present in red blood cells, to form oxyhaemoglobin. It is a reversible reaction as oxyhaemoglobin can also release oxygen so that the oxygen is available to cells and tissues in the body. This can be represented by a simple equation:

haemoglobin + oxygen   oxyhaemoglobin

Each gram of haemoglobin can combine with 1.34 cm³ of oxygen, so the total amount of oxygen that can be carried in the blood depends mainly on the amount of haemoglobin present.

We measure the availability of oxygen (or other gases) in terms of its **partial pressure**, which can be considered to be equivalent to the oxygen concentration. If haemoglobin is exposed to progressively increasing partial pressures of oxygen, the haemoglobin molecules gradually take up oxygen until they are all fully saturated with oxygen. Each haemoglobin molecule can combine with up to four molecules of oxygen. The relationship between the partial pressure of oxygen and the quantity of oxygen combined with haemoglobin (measured as the percentage saturation of haemoglobin with oxygen) is an S-shaped curve. This is referred to as the **oxygen dissociation curve**. Figure 2.26 shows how the percentage saturation of haemoglobin with oxygen increases as the partial pressure of oxygen increases.

The exact shape of the dissociation curve depends on a number of factors, including the partial pressure of carbon dioxide, temperature, and pH. For example, if the partial pressure of carbon dioxide increases, the curve moves to the right. This is known as the **Bohr effect** and is due to a decrease in the affinity of haemoglobin for oxygen. In other words, oxygen is more readily released from the oxyhaemoglobin. The Bohr effect is important in tissues that are respiring rapidly. Respiration produces carbon dioxide so the partial pressure of carbon dioxide increases. This causes oxyhaemoglobin to release its oxygen more readily, so that more oxygen will be delivered to the tissues that need it for respiration. The effect of an increase in the partial pressure of carbon dioxide is shown in Figure 2.26.

Fetal haemoglobin (haemoglobin F) has a higher affinity for oxygen than adult haemoglobin, so fetal haemoglobin combines with oxygen more readily than adult haemoglobin does. This means that fetal haemoglobin will receive oxygen from the maternal oxyhaemoglobin, at the same partial pressure of oxygen. In this way, oxygen is transferred from the mother's blood to the fetal circulation, in the placenta.

Myoglobin is a pigment found in muscle tissue, particularly in the leg muscles and hearts of large mammals. Like haemoglobin, myoglobin can combine reversibly with oxygen, but myoglobin takes up oxygen much more readily than haemoglobin does. When blood reaches muscle tissue, oxygen is transferred from oxyhaemoglobin to the myoglobin, which acts as a temporary oxygen store in the muscles. Myoglobin releases oxygen when the partial pressure of oxygen in the muscle drops to a very low value, perhaps as a result of vigorous exercise.

Carbon dioxide, which is produced in respiration, diffuses from body tissues into the blood, where most of it is taken up by red blood cells. In the red blood cells, carbon dioxide combines with water to form carbonic acid (H_2CO_3). This reaction is catalysed by the enzyme carbonic anhydrase. The carbonic acid then dissociates forming hydrogencarbonate ions (HCO_3^-):

$$CO_2 + H_2O \rightarrow H_2CO_3 \rightarrow H^+ + HCO_3^-$$

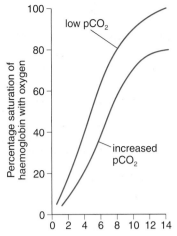

Figure 2.26 *The oxygen dissociation curve showing the effect of an increase in the partial pressure of carbon dioxide (pCO₂)*

QUESTION

Under what circumstances might the amount of haemoglobin present in the blood change?

TRANSPORT SYSTEMS

Table 2.1 *Cell types present in the blood of humans*

Cell type	Size / μm	Number per mm^3	Function
erythrocyte	6 to 8	4 to 6 million	transport of respiratory gases
leucocyte (neutrophil)	10 to 12	2800 to 5250	phagocytosis of pathogenic microorganisms
leucocyte (eosinophil)	10 to 12	70 to 420	secrete 'major basic protein' which is involved in defence against certain parasitic worms
leucocyte (basophil)	9 to 10	0 to 70	secrete heparin and histamine
leucocyte (lymphocyte)	7 to 8	1400 to 3150	secrete antibodies
leucocyte (monocyte)	14 to 17	140 to 700	migrate out of blood to form macrophages: phagocytic cells which engulf bacteria and cell debris
platelets (thrombocytes)	2 to 3	150 000 to 400 000	release thromboplastin, which is important in blood clotting

The hydrogencarbonate ions then diffuse out of the red blood cells into the plasma, in exchange for chloride ions (Cl^-), which move from the plasma into the red cells. The majority of the carbon dioxide is transported in the plasma in the form of hydrogencarbonate ions.

Carbon dioxide also reacts with haemoglobin and other proteins, to form compounds known as carbamino compounds. Carbon dioxide is relatively soluble in the plasma and a small percentage of carbon dioxide is transported in the form of a simple solution, dissolved in the plasma.

To summarise, carbon dioxide is transported by the blood in three ways:
- as hydrogencarbonate ions (HCO_3^-)
- as carbamino compounds
- in simple solution, dissolved in the plasma.

Interchange of materials between capillaries and tissue fluids

The walls of capillaries consist of a single layer of flattened epithelial cells, which act as a selectively permeable membrane, allowing water and solutes of low molecular mass to pass through. Proteins remain in the capillaries because their molecular mass is too great. The fluid which leaks out of the capillaries is called **tissue fluid**. It fills the spaces in between body cells. Tissue fluid contains water, glucose, amino acids, fatty acids, glycerol, mineral ions, dissolved gases and vitamins. Tissue fluid provides a medium for the transport of materials such as glucose, oxygen and carbon dioxide, amino acids and mineral ions between blood in the capillaries, and the surrounding body cells. (Figure 2.27).

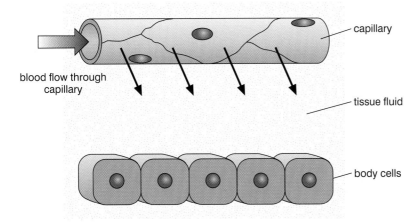

Figure 2.27 The formation of tissue fluid. Water and dissolved solutes, such as glucose and amino acids, are forced out of the capillary by blood pressure

At the arteriolar end of the capillaries, the blood pressure is relatively high (about 5.3 kPa) and this forces water out of the capillaries. However, the plasma proteins, which remain in the capillaries, exert an osmotic effect that tends to draw water back into the capillaries. The blood pressure at the arteriolar end of the capillaries is greater than the osmotic effect of the plasma proteins so the overall effect drives water out of the capillaries. At the venular end of the capillaries, the blood pressure has dropped to about 1.3 kPa. This is now lower than the osmotic effect of the proteins, which remains the same. There is, therefore, an overall inward pressure, which draws water back into the capillaries, by osmosis.

Tissue fluid is continually being formed at the arteriolar end of the capillaries and reabsorbed at the venular end. However, not all of the tissue fluid is reabsorbed in this way. Some drains into blind-ended lymphatic capillaries, which are similar to blood capillaries, except that their walls are more permeable (Figure 2.28). Excess tissue fluid, including some excretory products such as urea, enters the lymphatic system, where the fluid is known as **lymph**. This is eventually returned to the blood system via the lymphatic ducts.

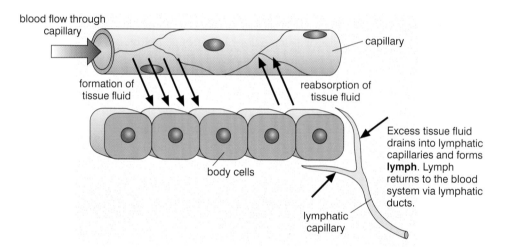

Figure 2.28 Formation and reabsorption of tissue fluid. The arrows show the direction of fluid flow

Adaptations to the environment

All organisms are adapted to survive in particular environmental conditions. Such adaptations may be:

- structural, as in the organism's body shape, or the presence or absence of appendages; or
- physiological, as in regulation of the composition of body fluids in response to the salt content of the environment.

Depending on the habitat in which they live, organisms may have a combination of structural and physiological adaptations.

Environmental factors affecting distribution and adaptation

Several physical factors affect the distribution of organisms in their habitats. These physical factors are often referred to as **abiotic**, to distinguish them from **biotic** factors, which involve the effects of other living organisms, including humans, on the distribution of species. Some physical factors, such as light, have widespread effects and are important in both aquatic and terrestrial situations. Others, such as wave action, are significant only in aquatic environments. The physical factors can be divided into:

- climatic – temperature, light, wind and water availability
- soil – often referred to as **edaphic** factors
- topographic – altitude, aspect (whether north-facing or south-facing), and inclination (steepness of slope)
- others, such as wave action, which are relevant in specific situations.

QUESTION

Why should topographic factors affect the distribution of organisms in a habitat?

In **terrestrial habitats**, light, temperature, soil type and the availability of water are important factors governing the distribution of the plants and animals. In sand dunes, for example, very dry conditions exist initially and the colonising organisms are adapted to lack of water and unstable soil, as well as to exposure to wind. The first colonisers are plants that are salt-tolerant and able to develop extensive root systems in order to remain in position. A typical plant found in such situations is marram grass (*Ammophila arenaria*), which has many adaptations to deal with the lack of water, exposure and shifting soil.

In **aquatic habitats**, such as a pond or a stream, organisms live in different places according to environmental factors. If the water is flowing, for example, this might mean that some organisms are attached to the pond or stream bottom or to rocks and stones, so that they remain in the same position. Others swim or move so that they remain in the most favourable conditions. There are many structural adaptations in both plants and animals associated with the effects of movements and currents in the water. Streamlining and the possession of fins in fish, flexible stems and the variations in leaf form in plants and the flattened shape of crustaceans are all examples of structural adaptations to the movement of the water in an aquatic habitat. Aquatic organisms also show structural adaptations to other factors, such as light, oxygen concentration and salinity.

ADAPTATIONS TO THE ENVIRONMENT

Structural adaptations

Species are adapted to survive in particular environmental conditions and some of their structural features are linked to the physical characteristics of their habitat. The distribution of flowering plants is affected by water availability and many are structurally adapted to live in dry habitats. Where plants are totally submerged in water, different structural adaptations enable survival in the conditions that prevail.

Xeromorphic adaptations in flowering plants

Any feature that significantly reduces the evaporation of water from the aerial parts of a flowering plant can be considered to be a **xeromorphic** adaptation. Most of a plant's water loss occurs during transpiration (as water vapour diffuses out through the open stomata on the leaves), although some water is lost by evaporation through the cuticle. The rate of transpiration is affected by temperature, light intensity, humidity and air movements.

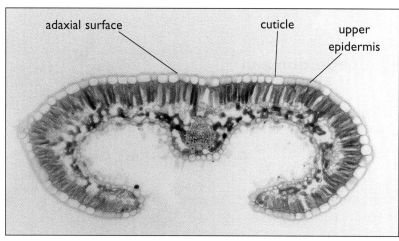

Figure 3.1 Xeromorphic adaptation in heather (Erica sp.) reduces water loss through leaves

Xeromorphic adaptations present in flowering plants include:

- thicker cuticles on leaves and herbaceous stems (Figure 3.1) – for example, on the upper (adaxial) epidermis of leaves of heather (*Erica* sp.) and holly (*Ilex* sp.) but also on the lower (abaxial) epidermis of marram grass.

- reduction in the size of leaves – there often is a reduction in the surface area to volume ratio, resulting in a decrease in the area of the leaf blade where most of the stomata are situated. There may be a corresponding increase in the thickness of the leaf blade. For example, heather has small leaves whilst in other plants, such as gorse and broom, the leaves are reduced to spines and photosynthesis takes place in the green stems.

- curling or rolling of the leaves into a cylindrical shape, to reduce the surface area of the leaves – for example, in marram grass, leaf rolling encloses the upper epidermis, where the stomata are situated. The lower epidermis has a thicker, waxy cuticle and no stomata.

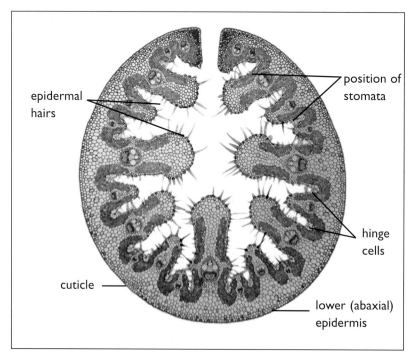

Figure 3.2 Several adaptations in marram grass (Ammophila arenaria) reduce water loss

Humid air is trapped inside the rolled-up leaf, reducing the diffusion gradient and hence the rate of transpiration. Specialised epidermal cells, known as *hinge cells*, are present on the upper epidermis and lose water rapidly when the transpiration rate is high, causing the leaf to roll up (Figure 3.2).
- the number and distribution of the stomata – stomata may be confined to pits or grooves on the underside of the leaves, so that humid air is trapped and transpiration decreases.
- the presence of epidermal hairs – these trap humid air and reduce the effects of air movement on the rate of transpiration.

In addition to these adaptations, there may be an increase in the quantity of supporting tissues, such as collenchyma and sclerenchyma. In habitats where water loss due to excessive transpiration is high, wilting may occur and extra supporting tissues can prevent the collapse of the herbaceous (non-woody) tissues, allowing time for the water balance to be restored. Many flowering plants lose their leaves during the season of the year when water is in short supply. These plants are known as **deciduous**. Evergreen plants retain their leaves throughout the year and often show xeromorphic adaptations, such as thicker cuticles.

Hydrophytes

Hydrophytes grow in situations where water is freely available, such as in and around ponds, lakes and streams. Some are adapted to being totally submerged, such as Canadian pond weed (*Elodea canadensis*) and water milfoil (*Myriophyllum spicatum*), whilst others may be rooted in the water with their stems, leaves and flowers projecting above the surface, such as water lilies (*Nymphaea alba*) and water crowfoot (*Ranunculus aquatilis*).

Totally submerged hydrophytes usually have:
- no cuticle (Figure 3.3)
- no stomata (Figure 3.3)
- reduced vascular and supporting tissues (Figure 3.3)
- small leaves or leaves with a dissected lamina (Figure 3.4)
- air spaces or air bladders (Figure 3.3)
- reduced root systems.

These adaptations enable the plants to carry out gaseous exchange efficiently, whilst offering little resistance to the movements of the water. Large expanses of leaf blade or rigid stems could be damaged by water movements, whereas air spaces and air bladders provide buoyancy, keeping the plants near the surface of the water and in the best position for efficient photosynthesis. Anchorage is not always of great importance and water uptake occurs all over the surface, so reduced root systems are common in totally submerged hydrophytes.

Some hydrophytes, such as water lilies, are rooted at the bottom of ponds and have leaves that float on the water surface (Figure 3.5). These leaves have several special adaptations including:
- stomata on the upper surface
- absence of stomata on the lower surface

QUESTION

From the text and what you can see in Figure 3.2, list all the structural xeromorphic adaptations shown by marram grass.

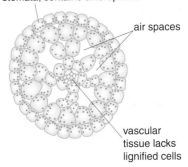

epidermis lacks cuticle and stomata; contains chloroplasts

air spaces

vascular tissue lacks lignified cells

*Figure 3.3 Transverse section through leaf of water milfoil (*Myriophyllum sp.*), showing adaptations in totally submerged hydrophyte.*

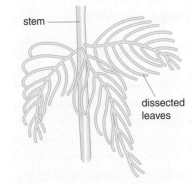

stem

dissected leaves

*Figure 3.4 Dissected leaves of water milfoil (*Myriophyllum sp.*) enable efficient gas exchange in an aquatic habitat*

ADAPTATIONS TO THE ENVIRONMENT

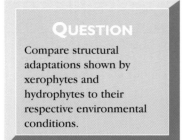

QUESTION

Compare structural adaptations shown by xerophytes and hydrophytes to their respective environmental conditions.

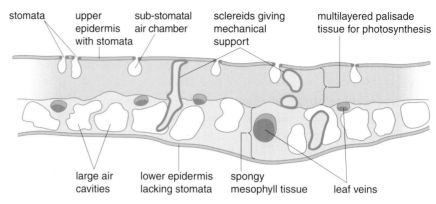

*Figure 3.5 Transverse section through a leaf of water lily (*Nymphaea *sp.). The upper epidermis consists of thin-walled cells with a very thin cuticle.*

*Figure 3.6 The aerial and submerged leaves of water crowfoot (*Ranunculus aquatilis*) have different adaptations because of their environment*

- elongated lignified cells, called *sclereids*, that span the leaf between the upper and lower epidermises and prevent the leaf from rolling up
- a thick palisade mesophyll layer.

In addition, the petioles (leaf stalks) help to keep the leaf floating on the surface of the water.

Some hydrophytes, such as water crowfoot, have both aerial and submerged leaves (Figure 3.6). The submerged leaves often resemble those of plants such as Canadian pond weed or the water milfoils, being either small or dissected, and lacking a cuticle and stomata. The aerial leaves resemble those of mesophytes, with broad leaf blades, cuticles on the epidermis, and stomata.

ADDITIONAL MATERIAL

Adaptations to salinity in flowering plants

Plants that can tolerate high levels of salt are referred to as **halophytes**, and typically are found growing in estuaries and salt-marshes, where their roots may be immersed in sea water. The degree of salinity to which they are exposed can vary, depending on their location and on the tides. Compared with sea water, salinity levels can be lower in a tidal estuary, but higher in a salt-marsh, because of the evaporation of water from the soil at low tides. Adaptations to these changing conditions often involve the maintenance of high salt concentrations in the plant tissues or the development of special tissues in which water can be stored.

Some animals can adapt to changes in salinity but, as many animals are motile, they generally will move to an area where they are best suited to the prevailing conditions.

Adaptations to water movements in aquatic organisms

Water movements include currents, and the ebb and flow of tides (the latter possibly interacting with air movements to bring about wave action). Any movement of water has an eroding action on soil and rocks and moves living organisms from one place to another, unless they are rooted or attached to a substratum. The churning action of water creates aeration of the water and also may contribute to the water's turbidity. Organisms in aquatic habitats show a wide range of adaptations to life in water.

Rate of flow in streams or rivers is an important parameter because of its influence on the organisms inhabiting the water. As the current increases, organisms that are unable to swim against it, or to take hold, are likely to be washed away. Faster flowing water is likely to be better oxygenated than sluggish or still water because of the mixing effect.

Adaptations to changing oxygen concentration as shown by invertebrates in fresh water

A supply of oxygen is vital for all organisms that respire aerobically in any habitat. The oxygen content of the atmosphere is more or less constant at around 21 per cent by volume. Oxygen content of the soil atmosphere is slightly lower, due to the respiration of soil organisms. In aquatic habitats, oxygen concentration can be very variable. In still, undisturbed water, the oxygen content may be very low, with very little, if any, oxygen (anaerobic conditions) in the mud at the bottom. Any disturbance of the water brings about aeration, so the water in a fast-flowing stream has a greater oxygen concentration than that in a pond. Also, the presence of actively photosynthesising plants and algae in the water can add significantly to the oxygen content, particularly on warm, bright days.

Oxygen is not very soluble in water so it is less available to aquatic organisms than to terrestrial organisms. In addition, as the temperature increases, the solubility of oxygen decreases. At 30 °C, the oxygen content of water is about one half of that at 0 °C (Figure 3.7).

Another factor to consider is that oxygen diffuses faster in air than it does in water. At 20 °C, the diffusion constant for oxygen in air is 11.0, whereas it is only 0.000034 in water. As air is less dense than water, convection currents are more easily established in air. However, ventilation movements associated with gaseous exchange require less energy.

There are many ways in which animals living in fresh water obtain the oxygen they require, and there are several adaptations to the varying oxygen concentrations found in aquatic habitats. Amongst invertebrates, oxygen may be obtained by diffusion over the entire body surface, or via specialised respiratory structures, such as gills, tracheae or lungs. The presence of respiratory pigments and circulatory systems enhances the distribution of oxygen from the respiratory surfaces to the metabolising cells.

In the smallest aquatic invertebrates there is a large surface area to volume ratio and the passive uptake of oxygen by diffusion from the surrounding water is sufficient to supply their needs. In *Amoeba* sp. (Figure 3.8a), there is a lower concentration of oxygen inside the organism, due to aerobic

QUESTION

Amoeba and *Hydra* do not have external gills or respiratory pigments. Suggest how these organisms are able to obtain sufficient oxygen for their needs.

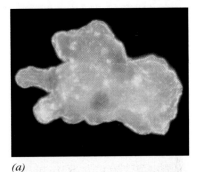

(a)

(b)

Figure 3.8 Two aquatic organisms: (a) Amoeba *and (b)* Hydra

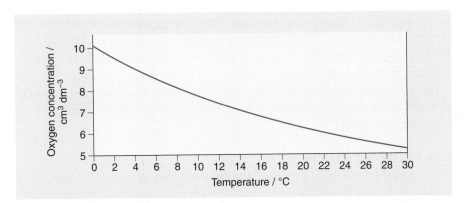

Figure 3.7 The dissolved oxygen content of water decreases as water temperature increases

respiration during metabolism, and a higher concentration of oxygen in the surrounding water. A diffusion gradient therefore exists. However, high metabolic rates can be maintained only if such organisms are 1 mm or less in diameter. The rate of diffusion through an organism decreases as the bulk increases.

In *Hydra* sp. (Figure 3.8b), both body layers (the ectoderm and endoderm) are thin: the inner endoderm layer is bathed by the water that circulates around the enteron (body cavity), and gaseous exchange can occur here as well as in the outer ectoderm layer. In water fleas, such as *Daphnia*, oxygen diffuses in over the entire body surface when the oxygen concentration is high. When the oxygen content of the water is low, *Daphnia* develop relatively high concentrations of haemoglobin to compensate for the reduction. This gives the organisms a pink colour.

Bulkier invertebrates, such as the flatworms (Platyhelminthes), which lack specialised respiratory structures and circulatory systems, have flattened bodies, thus increasing their surface area to volume ratio and reducing the internal distance over which diffusion takes place.

Circulatory systems and respiratory pigments

Annelids, such as the oligochaetes (e.g. earthworms), polychaetes (e.g. ragworms) and leeches, have circulatory systems, although gaseous exchange takes place over their entire body surface. The surface area to volume ratio enables sufficient oxygen to be taken up and steep diffusion gradients are maintained. The oxygen quickly diffuses into the capillaries below the epidermis and is transported away to other parts of the body by the blood system. The respiratory pigment, haemoglobin, is present in the blood and enhances the oxygen-carrying capacity of the blood. This enables oxygen to be picked up readily in situations where it is more abundant and released where it is needed.

Tubifex is an oligochaete (Figure 3.9), which is well adapted to aquatic habitats where the oxygen concentration is in short supply. It lives, head downwards, in the mud at the bottom of rivers. Here anaerobic conditions prevail as a result of oxygen depletion due to the activities of aerobic bacteria following high levels of organic pollution. Oxygen is obtained by diffusion through the posterior body wall and is taken up by the respiratory pigment, haemoglobin, present in the blood. *Tubifex* wave their tails rhythmically in the water: the lower the concentration of oxygen in the water, the greater the rate of tail-waving.

Respiratory pigments are also present in the larvae of some of the members of the genus *Chironomus*. These larvae are known as 'blood worms' (Figure 3.10) and are able to survive in conditions where the oxygen concentration is low. In addition to the respiratory pigment, these larvae have gills, which are extensions of the body through which the blood flows and which are in contact with the water. These gills increase the surface area over which gaseous exchange can occur.

Gas exchange structures

Adult insects have a well-developed **tracheal system**. Tracheae are tubes formed by ingrowth of a layer of cells and are lined with cuticle, thickened to form rings or spiral ridges and thus keeping the tubes open for the passage of

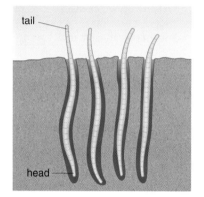

Figure 3.9 Tubifex *sp. obtain oxygen by diffusion; the lower the oxygen concentration in the water, the greater the rate of tail-waving*

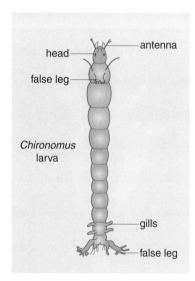

Figure 3.10 External gills in Chironomus *increase the surface area over which gaseous exchange can occur*

gases. Typically, these tubes open to the outside by means of **spiracles**, located on the thorax and abdomen. In fully aquatic insects and the aquatic larvae of insects, such as dragonflies and mayflies, spiracles are absent and so-called '**tracheal gills**' are present. These are plate-like structures, containing many fine branches of the tracheal system, and are situated immediately below the cuticle. In some dragonfly nymphs, these tracheal gills are situated in the rectum and are ventilated by muscular movements that also contribute to the locomotion of the nymph. There is no circulatory system or respiratory pigment involved in gaseous exchange in these organisms.

Aquatic insects that live in still water rely on the atmosphere for their oxygen supplies. These insects are able to rise to the water surface, in order to obtain air through spiracles, without being washed away by water currents. Adaptations such as **breathing tubes** or **siphons** (Figure 3.11) are found in the larvae of gnats (*Culex pipiens*) and drone flies (*Eristalis* sp.). In the gnat larvae, the siphons pierce the surface layer, whereas in the drone fly larvae (known as 'rat-tailed maggots'), which live on the bottom of shallow bodies of water, the siphons are telescopic and can be extended to 6 cm or more. The openings of the siphons are often surrounded by a fringe of hairs to prevent waterlogging.

A number of beetles and water bugs transport air bubbles down into the water. The water boatman, *Notonecta glauca* (Figure 3.12), traps a single bubble among the hairs at the base of the abdomen. When this air has been used, the insect returns to the surface to obtain a fresh supply. In this way, sufficient oxygen is obtained for the organism's needs.

Common pond snails, such as *Limnaea* and *Planorbis*, are adapted to life in still water with varying oxygen concentrations. In these molluscs, the inner surface of the mantle cavity (gas exchange surface) functions as a **lung**, and they come to the surface to replenish their air supplies. The frequency with which they need to do this depends on the oxygen levels in the water. *Planorbis* has haemoglobin in its blood and thus is better adapted to poorly oxygenated conditions than *Limnaea*. The haemoglobin acts as an oxygen store and enables the animal to remain submerged for long periods of time. In contrast, *Limnaea* has no haemoglobin and the oxygen content of its blood decreases quite quickly, with a consequent decrease in its metabolic activity, so it must return to the surface more frequently.

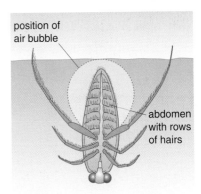

*Figure 3.12 The water boatman (*Notonecta glauca*) obtains oxygen by trapping bubbles of air under its abdomen*

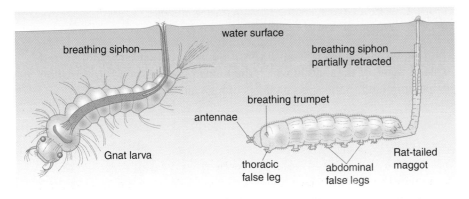

Figure 3.11 Different adaptations in these larvae allow for gas exchange

4 Human ecology

Figure 4.1 Humans live in and tolerate a wide range of physical conditions, from the heat and aridity of the desert in Egypt (top) to the sub-zero temperatures of the North West Territories of Canada (bottom)

The human species – distribution and tolerances

Human beings (*Homo sapiens*) are remarkably versatile in their living or habitat tolerances. Other animal species are often quite precise in their niche or habitat requirements and this is reflected in their distribution on a local and broader geographical scale. If we looked at a wide spectrum of animal species we could identify quite precise limits to their tolerance of a range of physical factors, such as light, temperature, altitude, salinity, oxygen concentration and pH.

Humans exist in a wide range of terrestrial environments, having extended their territories by use of buildings and clothing that protect them from unsuitable or extreme environmental conditions (Figure 4.1). Historically, regions inhabited by human populations have been determined largely by accessibility to land for food production and by the need for water – for people, for their crops and for their animals. Nevertheless, human populations show tolerance to wide temperature ranges, for example, from a hot 50 °C in July in the Sahara in Africa to a cold −65 °C in January in Yakutsk (in the former USSR). Most, however, prefer to live well away from these extremes. The upper limit for permanent settlements at high altitude is about 4500 m and people sometimes go a short depth underground in mines, in road and rail tunnels or under cities.

The human species has thus colonised a wide range of habitats and has shown its ability to tolerate considerable variation in the physical factors of the external environment. Remember, however, that many people spend a relatively large amount of their time in artificially modified environments – in houses, shops and offices. Nevertheless, the **internal environment** of the human body is maintained within quite narrow limits. The ability of an organism to control its internal environment is referred to as **homeostasis**. This is the result of internal physiological mechanisms that regulate factors such as body temperature, body water content, ionic composition, blood glucose concentration and oxygen concentration in the blood. The areas populated by the human species and the behavioural patterns that have evolved are such that the body regulatory mechanisms can operate satisfactorily.

This chapter on human ecology looks at humans in their environment and explores some of the ways in which the human body responds to variations in temperature and to the special effects of high altitude.

BACKGROUND MATERIAL

Beyond the limits and some exceptional feats

In terms of altitude, while the upper limit for permanent settlements on land is about 4500 m, inside aircraft people are now able to cruise regularly at 10 000 m. Space travel has taken people to the moon, introducing space travellers to the experience of weightlessness. Travel in both aircraft and spacecraft is possible only by creating an artificial environment inside.

Figure 4.2 Ranulph Fiennes and Mike Stroud completing their 2050 km journey, on foot, across the Antarctic – an example of an exceptional feat

Occasional exceptional feats have taken a few individuals outside the normally inhabited areas, but if the adverse conditions are too extreme or the human is subjected to them for too long, there comes a point when the body can cope no longer. We can look at the endurance shown by Ranulph Fiennes and Mike Stroud in the Antarctic during the winter of 1992 to 1993. For 97 days they journeyed on foot, dragging all their own supplies and equipment in their attempt to walk across the Antarctic continent (Figure 4.2). We can contrast their success with the failure of the Antarctic expedition in 1912, led by Robert Scott. One by one the men failed physically or mentally and though they reached the South Pole, none survived. All were beaten by starvation, cold and frostbite before they could reach safety and the ship home.

QUESTION

The estimated average energy requirement for a man aged between 19 and 50 years is about 20.6 MJ per day (1 MJ = 1000 kJ). The energy expenditure of Dr Mike Stroud and Sir Ranulph Fiennes, during their expedition across Antarctica, was between 23 and 44 MJ per day. What sort of diet would be required to provide this energy?

Temperature

Body temperature

Human beings are described as **endotherms**. This means that, like other mammals, humans are able to maintain a constant high body temperature independently of the external environmental temperature. In humans, normal body temperature is about 37 °C. With a clinical thermometer, we usually take the oral temperature (in the mouth) but we can also take a person's rectal temperature. The two values would be slightly different. The oral temperature is less reliable as it may vary according to recently consumed hot or cold foods and drinks, or because of breathing activities. If we were to use a thermocouple to measure temperature with greater precision, we would find that the body temperature is not the same throughout the tissues. The rectal

Figure 4.3 Diurnal variation in body temperature: mean body temperature of a group of 14 men, measured over a period of 4 days in a temperate climate. All were carrying out the same task.

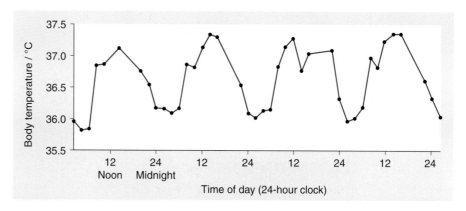

temperature is close to the temperature of the deeper structures in the body, or the body **core**. Thus, when we speak of a constant body temperature, strictly speaking the term refers to the core temperature of the body, that is, the temperature inside the head, thorax and abdomen.

In humans, the core temperature fluctuates within a narrow range, normally between 36 and 37.5 °C. There is a clear **diurnal rhythm**, with daily temperatures reaching a peak from midday to early afternoon and a trough during the night after midnight (Figure 4.3). This rhythm gradually becomes reversed if daily activity patterns change, for example, in workers on night shifts. In women the temperature is lowest during menstruation and rises noticeably at about the time of ovulation. Exercise may lead to a rise in core temperature, with a temperature as high as 41 °C being recorded in an athlete after a marathon race. Core temperature may also rise in response to emotional situations.

Temperature regulation

If a steady temperature is to be maintained in the human body, heat loss must equal heat gain (Figure 4.4).

Figure 4.4 The balance between heat gain and heat loss in the maintenance of a stable body (core) temperature. Heat gain may be the result of internal generation (thermogenesis) from metabolism, including exercise and shivering, or external gain from the environment, hot food and drink. Heat loss involves the skin, through conduction, convection, radiation and evaporation through sweating.

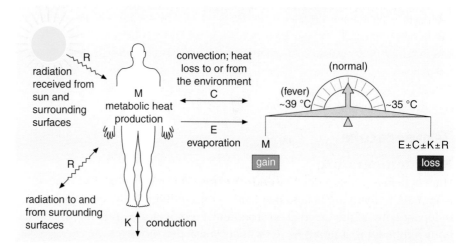

Heat gain – Most of the heat gained is generated internally from metabolic activities (thermogenesis), though there may be some gain from the surroundings. Heat is produced at a fairly constant rate from metabolic activities in various organs in the body, such as the liver and heart. For a person at rest, the heat produced is approximately 4 kJ per kg body mass per hour. This is equivalent to about 170 kJ per square metre per hour, or about the same as a 40 watt light bulb. Activity in skeletal muscles increases heat production and short bursts of vigorous exercise may increase the heat produced by more than 10 times the level at rest. A brisk walk (or jog or game of squash) does a lot to warm the body on a cold day. Shivering is a specialised form of uncoordinated muscle activity that produces heat, to about five times the level at rest. Shivering thermogenesis may be initiated when the body becomes cooled and lasts for a few minutes at a time.

Heat may be gained from the environment in situations where air temperature is higher than the temperature of the skin. The body also gains heat by direct radiation from the sun, from a fire or artificial heater, and indirectly from reflected radiation. Some heat is gained directly from the consumption of hot food and drink.

Heat loss – When the body is hotter than the surrounding environment, heat may be lost by direct radiation from the body, by conduction through areas of the body touching cooler objects, by convection to the surrounding air or by evaporation of water in sweat. The skin covers the surface of the body and plays a very important part in regulating the loss of heat from the body and thus in maintaining the required balance between heat gain and heat loss.

The role of skin in temperature regulation
To understand the role of the skin in temperature regulation, it is important to look at its general structure, shown in Figure 4.5. Heat loss from the skin occurs by conduction, radiation and evaporation of sweat. Heat loss by conduction and radiation can be varied by altering the blood flow to superficial capillaries. There are blood vessels that can direct blood flow either to superficial capillaries or across to veins. These are known as **shunt vessels**.

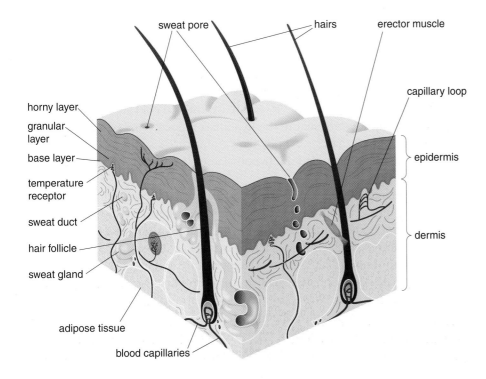

Figure 4.5 The structure of human skin showing structures involved in temperature regulation

The diameter of arterioles in the skin is controlled by sympathetic nerves originating in the hypothalamus in the brain. External cold results in a decrease in the diameter of these vessels. This is known as **vasoconstriction**, and results in a *reduced* blood flow to capillaries in the skin. Conversely, **vasodilation** (an increase in the diameter of skin arterioles) occurs in warm conditions. This results in an *increase* in the blood flow in the peripheral circulation and leads to a considerable increase in heat loss.

In hot situations, an important means of heat loss is through evaporation of water from the sweat glands or from the inside surface of the lungs and mouth. When 1 g of water vapourises it requires 2.42 kJ of heat and this latent heat is taken from the immediate surroundings, resulting in cooling. 'Thermal' sweating occurs when there is an increase in external temperature or a rise in body temperature. (Note that sweating may occur in conditions other than increased body temperature – for example, 'mental' sweating occurs in response to emotional states.) Sweat glands secrete a dilute solution containing salt (sodium chloride), urea and lactic acid. Heavy sweating involves rapid loss of water and salt from the body. Dehydration and salt deprivation will occur unless enough water is drunk and adequate amounts of salt are taken to replace the losses.

In cold conditions, smooth muscles attached to hair follicles in the skin contract, raising the hairs. This is important in many mammals as it helps to trap a layer of air in their fur. Air is a poor conductor of heat so this helps to insulate the body against excessive heat loss. The insulating effect of hair is minimal in humans, except on the head, and lack of hair in bald adults and babies can lead to significant heat losses. We should all take heed of the advice to wear a hat in cold weather! The layer of adipose tissue also makes a contribution as a means of insulation because fat is a poor conductor of heat.

Temperature receptors and the role of the hypothalamus

The skin contains special heat-sensitive nerve endings, known as **thermoreceptors**, or temperature receptors. These send nerve impulses to the spinal cord and then to the hypothalamus in the brain. If the temperature of the skin changes, for example if you are exposed to cold air, the rate at which nerve impulses are sent to the hypothalamus also changes. Nerve endings sensitive to changes in body temperature are also present in the spinal cord, the abdomen and the hypothalamus itself.

The hypothalamus therefore receives information about body temperature from several different sources and acts rather like a thermostat to maintain a nearly constant body temperature. The hypothalamus contains a temperature regulating centre, which compares information from the temperature receptors with the so-called 'set-point' temperature and, if there is a difference, makes appropriate responses to control heat loss or heat production by the body. As an example, if the thermoreceptors indicate a slight fall in body temperature, the hypothalamus initiates changes such as a decrease in blood flow in the skin capillaries, thus reducing heat loss from the skin by radiation. This helps to keep body temperature nearly constant.

Similarly, if body temperature rises, the hypothalamus initiates changes such as an increase in the rate of sweating. This leads to an increase in heat loss from the body as a result of the evaporation of sweat.

The temperature regulating centre in the hypothalamus is very sensitive to changes in body temperature – a change of less than 0.1 °C can bring about appropriate responses to maintain body temperature.

Behavioural mechanisms of temperature regulation

In most situations people wear clothes. This clothing has a protective effect which alters the exchange of heat between the body and the immediate environment. Clothes have an insulating effect when dry, so, to some extent, clothes modify the relationship of the skin with the immediate environment by creating a 'microclimate' close to the body. We spend a considerable proportion of our time indoors and the design of buildings also serves to protect us from the natural environmental temperature. Clothing and buildings are behavioural responses which help maintain the required balance between heat loss and heat gain. Details of clothing and buildings adopted by both natives and visitors in hot and in cold climates are described in the relevant sections later in this chapter (see page 65 and pages 66 to 68).

Extremes of temperature

The next sections look at how humans respond to extremes in environmental temperatures in both hot and cold environments. We describe the various physiological and behavioural responses that the body makes, but when conditions are such that temperature regulation processes are inadequate or fail, the body suffers from stress.

Response to high temperature and acclimatisation

If you live in a cool or temperate climate and pay a visit to a much hotter place, in the tropics or hot desert, at first you feel very uncomfortable. For a few days you are likely to lack energy even for walking around. Perhaps you find you cannot think very clearly and probably have little appetite. Then you begin to adjust, or **acclimatise**, and gradually find you can carry on your daily activities much as you did in the cooler place. In response, you would be wise to start your activities early in the morning when it is cooler, and be less active around midday and early afternoon when temperatures are at their highest. It would also be sensible to wear loose and generally light-coloured clothing, but you should not necessarily abandon all clothing and you should certainly not sunbathe in the direct heat of the sun.

In the hot climate, reduction of physical activity and reduced food intake mean that less heat is generated internally as a result of muscular and metabolic activity. A preference may develop for foods with high water content, such as fruit and salads. This may be a response of thirst mechanisms in relation to the potential threat of dehydration in the heat. In response to the high temperatures, blood vessels in the skin dilate, bringing more blood near the surface. But because the environmental temperature may be close to or higher than normal body temperature, there is likely to be little or no heat loss to the

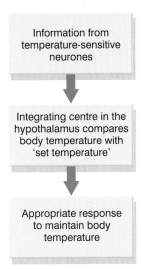

Figure 4.6 Diagram to show the central processes involved in the regulation of body temperature

surroundings as a result of conduction, convection or radiation. Evaporation of sweat thus becomes the most important means of losing heat and so cooling the body.

Sweating and sweat production – salt loss and dehydration

We will consider the effects of sweating in two different extremes of hot climate: hot and humid (as in tropical regions or in factories, mines and other enclosed environments) and hot and dry (as in desert conditions). When humidity is high, less sweat evaporates and sometimes sweating may cease altogether. Inability to lose heat through evaporation of sweat in these situations is serious and quickly leads to rise of the body core temperature. At the other extreme, if humidity is low, sweating is an effective way of losing heat. In temperatures as high as 40 °C a person may feel perfectly cool and comfortable, particularly if the clothing is loose, thus allowing sweat to evaporate from the body surface. The danger comes after prolonged exposure to high temperatures if the components of sweat are not replaced. Excessive loss of water through sweating can lead to severe dehydration of body tissues. Loss of ions in the sweat, particularly of sodium (Na^+) and chloride (Cl^-), depletes these from body fluids and is likely to lead to painful muscular cramps.

Increase in sweat production appears to be the main response to hot conditions as the body acclimatises. This effect is illustrated in the Extension material, in Figure 4.7. Data for this study were obtained from people doing physical work in an artificial environment, who were subjected to high temperatures for different lengths of time. Equivalent results are obtained when the body temperature is raised without the person indulging in physical work.

It seems that similar increases in the rate of sweating occur in people of different ethnic groups, though females usually show lower rates than males. This may be because females tend to undertake less vigorous physical activity than men, though it is of interest that some female athletes do show rates of sweating which are comparable with those of men.

People who normally live in hot climates appear to have an initially high rate of sweating compared with those from cooler climates but show a similar response when exposed to higher than normal temperatures. There is evidence to suggest that people born in 'hotter' climates and who are thus exposed to high temperatures as infants show increased activity of their sweat glands. As adults, these people have better tolerance to heat compared with people from cooler climates. In addition, the body core temperature falls (perhaps by 1 or 2 °C). This means the mechanisms that cause an increase in sweating start to come into effect at a lower temperature.

Another effect of acclimatisation is that the sweat produced has a lower salt concentration, thus reducing the harmful loss of salt from body fluids. People living in hot climates may find it advisable (and desirable) to take extra salt with their food or drinks.

To summarise, the main features of acclimatisation to heat are an increase in sweat production and a decrease in the level of salt excreted in the sweat. These

EXTENSION MATERIAL

Acclimatisation and sweating

Figure 4.7 shows the results of a study of sweat loss during acclimatisation. You can see that the people showed a noticeable increase in sweat loss. The dashed line on the graph shows the average sweat loss of the people in the group of approximately 4 weeks before the experiment.

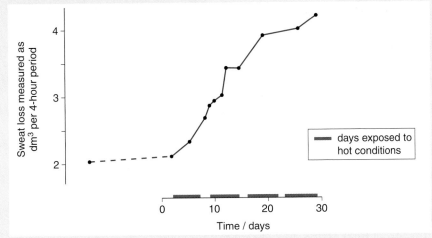

Figure 4.7 *A group of people exposed to hot conditions for 5 or 6 days at a time over a period of 4 weeks*

features of acclimatisation are effective probably over a relatively short time-scale and are rapidly lost if the person is no longer exposed to the extreme temperature. Adaptations shown by natives are not easy to define clearly. Their tolerance to heat, as for individuals experiencing acclimatisation, is helped by an increased sweat production and lower salt loss in sweat. Some natives in hot climates have the advantage of long limbs and arms and this is a means of exposing a relatively large surface area from which sweat can evaporate. However, not all people native to hot regions have the benefit of this body shape. Certainly, behavioural adaptations are probably of considerable importance and have become part of the traditional lifestyle. These include an early start to the day to coincide with cool temperatures, a relatively lethargic approach to activities (and often sleeping) during the hottest times of day, sensible use of shade and suitable clothing and buildings. These, and other, behavioural mechanisms are described in more detail elsewhere in this chapter.

Clothing and housing – People living in hot climates have evolved traditional clothing and housing that are usually highly effective in helping them to tolerate heat. Clothing that covers the body minimises direct exposure of the skin to the radiation of the sun; loose clothing allows for evaporation of sweat and also keeps a still layer of air next to the skin, which insulates against excessive absorption of heat; light colours reflect the heat (Figure 4.8). Clusters of compact, thick-walled mud houses with only narrow slits for windows, characteristic of villages in deserts, remain remarkably cool inside (Figure 4.9). Modern buildings use air conditioning, with controlled temperature and ventilation, in an attempt to provide comfortable working and living conditions, particularly for visitors from a cold climate.

Figure 4.8 *Traditional clothing of desert people: Bedouin in Sahara Desert*

Figure 4.9 *Typical desert village in Afghanistan, with thick-walled mud houses*

Heat stress – causes and effects

When the body can no longer cope with excessively high temperatures, problems of **heat stress** arise. When exposed directly to excessive radiation from the sun the skin may suffer from sunburn. Redness, blistering and peeling of the epithelium occurs. Increased production of the pigment melanin, associated with a tan, gradually gives some protection from the sun's radiation. An irritation known as prickly heat may develop due to blocking or narrowing of the ducts from the sweat glands, with the result that sweat cannot reach the skin surface.

In some situations a person may feel dizzy or suffer from **heat collapse**, probably because blood is diverted away from the internal organs to the skin and skeletal muscles. Recovery usually occurs quickly if the person is taken to a cool environment or lies down. **Heat exhaustion** is a more serious condition which may be due to dehydration because fluid lost in sweating has not been replaced. With a loss of 5 to 8 per cent of the body fluid, fatigue sets in, and with a 10 per cent reduction in body fluid, mental and physical deterioration occur. With further decrease in body fluid, in cases of severe dehydration, intracellular fluid is withdrawn and this starts to cause damage to cells. Heat exhaustion may also result when losses of sodium chloride are not replaced. **Heat cramps**, noticed as a stiffness in the muscles (particularly in the arms and legs), are often associated with heat exhaustion. These cramps develop as a result of the excessive loss of salt (as Na^+ and Cl^- ions) in sweat.

The most severe condition is known as **heat stroke**, characterised by a breakdown of the temperature regulation mechanisms. This appears to be associated mainly with a failure or inadequacy of the sweating mechanism. At core temperatures of 42 °C or higher, irreversible damage to cells and proteins occurs and the person is likely to go into a coma. Unless immediate steps are taken to lower the body temperature, death may follow. It should be appreciated that the upper limit of tolerance to temperature is only a few degrees above normal core temperature.

Estimated wind speed / mph ↓	Actual thermometer reading / °C											
	10	5	−1	−7	−12	−18	−24	−29	−35	−40	−46	−51
	Equivalent temperature / °C											
Calm	10	5	−1	−7	−12	−18	−24	−29	−35	−40	−46	−51
5	9	3	−3	−9	−15	−21	−26	−32	−38	−44	−50	−56
10	5	−2	−9	−16	−23	−31	−36	−44	−50	−57	−64	−71
15	2	−6	−13	−21	−28	−36	−43	−50	−58	−66	−73	−81
20	0	−8	−16	−24	−32	−40	−48	−55	−64	−72	−80	−87
25	−1	−9	−18	−26	−34	−43	−51	−59	−67	−76	−84	−92
30	−2	−11	−19	−28	−36	−45	−53	−62	−71	−79	−88	−96
35	−3	−12	−20	−29	−38	−47	−55	−64	−73	−81	−90	−99
40	−4	−13	−21	−30	−39	−48	−57	−66	−74	−83	−92	−101

Windspeeds greater than 40 mph have little added effect

↑ Little danger for properly clothed person, maximum danger of false sense of security

↑ Increasing danger from freezing of exposed flesh

↑ Great danger

Figure 4.10 The wind chill effect: the table shows how increasing wind speed lowers the effective temperature on exposed skin. Look for the wind speed in the left-hand column and the actual temperature across the top. From this, you can find the equivalent temperature on exposed skin. Wind speeds are given in miles per hour (mph) in line with familiar usage.

Response to low temperature and acclimatisation

The body's response to extreme cold is not simply a reverse of its response to heat. To begin with, 'coldness' may be due to very low environmental temperatures but people can also suffer from cold in more moderate temperatures, particularly in windy conditions. Moving air blows away the warm layer of air immediately surrounding the body and increases the loss of heat by convection. This **wind chill** factor (Figure 4.10) is made worse if the body surface and any clothing being worn are wet, because further cooling occurs due to evaporation of water.

Another difference when considering response to cold rather than heat is that the temperature gradient between the body surface and surrounding temperature is likely to be far greater in cold conditions. In many parts of the world, the surrounding temperature is well below the core temperature of 37 °C; indeed temperatures in the region of −40 °C may be experienced in some places normally inhabited by humans.

The body can tolerate and recover from a lowering of core temperature by as much as 10 °C, whereas a rise in temperature of only 4 °C may prove lethal. If the body temperature is to be maintained close to 37 °C, the focus is on generation of internal heat and on protection and insulation of the body as ways of **conserving** heat.

Increased metabolic rate in response to cold conditions leads to increased heat production. This can be stimulated by the hormones adrenaline and thyroxine. A person is likely to have a greater appetite for food in cold conditions, thus increasing food intake. Voluntary muscular movements during physical activity and involuntary shivering also contribute to the generation of heat, though shivering is likely to be sustained only for short periods.

EXTENSION MATERIAL

The wind chill effect and potential dangers

Here is an example of a 'calculation' using the wind chill effect table (Figure 4.10) that should help you to understand how to interpret data in the table. In an external temperature of −12 °C and with a wind speed of 30 mph, the surrounding temperature effectively becomes −36 °C. Even at 5 °C, a wind speed of 30 mph brings the effective temperature down to −11 °C. This shows how, in not remarkably cold temperatures, you can be exposed to conditions that are potentially dangerous to the body unless you have adequate protective clothing.

Diversion of blood away from the body surface by **vasoconstriction** is an important means of conserving heat. Reduced blood flow to the skin means that less heat is lost by conduction and convection from the body surface. Vasoconstriction for a prolonged period would restrict blood flow to the extremities of the limbs, that is, the fingers and hands, toes and feet. This could cause damage to the tissues but is counteracted by occasional dilation of the blood vessels in what is described as the 'hunting reaction'. Without this reaction, fingers tend to become numb or paralysed and loss of manual dexterity would lead to serious difficulties. The tendency to curl up the fingers, or indeed the whole body, with arms folded, when it is cold reduces the surface area exposed and illustrates another way of reducing heat loss to the surroundings.

A subcutaneous layer of fat provides **insulation**. This cannot be turned on and off at short notice, but development of a fat layer in certain people may be an advantage, for example to long-distance swimmers immersed in water for long periods. Young babies, up to about 6 months old, may benefit from 'brown fat', which generates additional heat by its metabolism rather than insulation effects. It is unlikely that brown fat plays a major part in the heat balance of adults.

Probably the most important way of protecting people from the cold is by conserving heat through **insulation** of the body, hence the importance of suitable clothing and buildings. In extreme temperatures and where wind chill might be significant, it is necessary for clothing to be windproof. Even though windproofing is important it can lead to discomfort due to condensation of moisture produced from the body. In extreme cold, this moisture may freeze inside the garment. Air is a good insulator, so the most effective clothing includes a number of layers which trap air, provided the air is dry (Figure 4.11). The design of clothing needs to allow for adequate physical activity and in this respect very heavy clothing may be cumbersome. In temperatures below freezing, the face and limb extremities become vulnerable if inadequately protected. Inside buildings additional heat is supplied artificially from fires, electrically or by other systems of central heating. An important economic consideration is the adequate insulation of the building as a means of saving energy and this is given high priority in the design of modern buildings. Traditional housing in cold climates usually has thick walls and small windows, thus minimising loss of heat, which is likely to be supplied from an open fire. The hard-packed snow used to construct Inuit (Eskimo) igloos has excellent insulating properties.

Figure 4.11 Clothing in an environment of extreme cold: an Inuit in the Arctic region of Canada

People living permanently in cold climates show certain physiological adaptations, though their main means of protection is through clothing and housing. One study showed that native Australians were able to sleep well, even though naked, in temperatures around freezing, whereas unacclimatised white people shivered violently and were unable to sleep properly in similar conditions. Inuit people have a greater blood flow through the hands and feet than do visitors in the same conditions, and the traditionally high protein diet of the Inuit may contribute to their high metabolic rate. Acclimatisation to cold by people who travel from warmer to colder climates does occur, but compared with acclimatisation to heat, it is less easy to define the mechanisms.

It is likely that increased food intake, leading to increased metabolic rate, is an important response, though the most important consideration is to ensure adequate protection by means of suitable clothing and housing.

Cold stress – causes and effects

When the body can no longer cope with the effects of cold, problems of **cold stress** may result.

Cold injury – Cold injury is characterised by actual damage to tissues, most often the hands, feet and face, because of their direct exposure to low temperatures. Severely reduced blood circulation deprives the tissues of nutrients and their normal metabolic reactions. In mild form, chilblains may develop as the parts become tender and itchy. In more severe conditions, tissues actually freeze and this is known as **frostbite** (Figure 4.12). If this is only superficial, the tissues are likely to recover and injured skin is replaced by new growth. If exposure is prolonged or tissues are deeply frozen, underlying tissues, including muscle and bone, may suffer permanent damage. This is caused by the mechanical action of ice crystals on the cell structure and also by the dehydrating effects of removing liquid water from body fluids. Gangrene may set in, resulting in loss of toes or fingers. The condition known as **trench foot** is usually a result of prolonged cooling in cold water. Trench foot is characterised by blackening of the skin of the toes and foot. The main damage is to the muscles and nerves and can lead to gangrene if the affected limbs are not warmed up quickly.

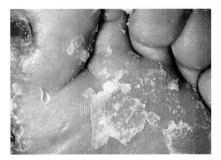

Figure 4.12 Frostbitten toes, after treatment. Frostbite refers to tissue damage due to freezing, from the destructive effects of the formation of extracellular ice crystals. Frostbitten parts need to be gently warmed in tepid water. Precautions are required against bacterial infection, to which frostbitten skin is susceptible

Hypothermia – Hypothermia develops when the core temperature falls to 35 °C or below. Down to 35 °C shivering increases, but already the muscles are likely to be at a lower temperature. Below 35 °C there are signs of muscle weakness and the person shows difficulty in walking and coordinating movements. By 34 °C the person becomes mentally confused and vision is disturbed. Loss of consciousness occurs between 32 and 30 °C and death usually follows at between 28 and 25 °C if the person is not warmed up.

Probably the most important effect of the lowering of core temperature is the reduction of the heart rate. Cooler blood flowing through the heart affects the **pacemaker** (sinuatrial node) that initiates the heart beat. Reduced blood output from the heart means that the coronary circulation and flow to muscles and the brain may be inadequate, hence the symptoms associated with hypothermia. People suffering from hypothermia may also show a slower respiration rate and increased production of urine. The latter is linked to suppressed release of antidiuretic hormone (ADH). However, recovery from hypothermia can be complete provided the body is warmed quickly.

A number of deaths from hypothermia, or **exposure**, do occur each year among walkers and mountain climbers, even on the hills of Britain, usually through a combination of physical fatigue and inadequate protective clothing. Reference to Figure 4.10 and the 'calculation' in the Extension material on page 67, emphasises how the wind chill factor can significantly lower the effective surrounding temperature and lead to situations that may result in hypothermia. Immersion in cold water can quickly lead to hypothermia and accidental deaths in water are often due to hypothermia rather than drowning. Compared with air, water has a higher thermal conductivity so the body cools faster when surrounded by water. Survival time of a naked unprotected person is about 90 minutes in water at 15 °C, but only 30 minutes at 5 °C. Attempts to swim to safety after an accident in cold water are probably misjudged, since the movement disturbs any remaining layer of warm water adjacent to the skin (that might have retained some warmth from the body) and the activity uses up valuable metabolic reserves.

Young babies and old people are particularly vulnerable to hypothermia. Up to about 1 year old, babies are unable to shiver and also their behavioural responses are limited. Old people generally show less physical activity and have a lower metabolic rate, thus reducing internal heat generation. Their temperature control mechanisms, originating in the hypothalamus, are usually less effective with respect to the responses such as shivering, vasoconstriction and increase in oxygen consumption. In some cases the hypothalamus sets the temperature at too low a level. For those living on low incomes the situation may be made worse by inadequate heating and clothing.

In some surgical operations, involving the heart or brain for example, hypothermia may be deliberately induced by cooling the blood or body surface. This is used as a short-term means of reducing blood flow and use of oxygen in the tissues and allows the operation to take place. In these situations, body temperature can be lowered temporarily to about 25 °C.

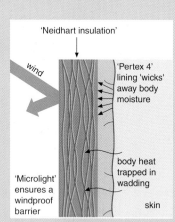

Figure 4.13 Clothing (from Rohan) designed to provide 'one stop insulation': an outer layer of 'Microlight', a middle layer of 'Neidhart insulation' (thin, stable, featherweight) and an inner layer of 'Pertex'. Successful insulation is achieved by trapping still air next to the body.

QUESTIONS

A question of clothing

A manufacturer of specialist outdoor clothing advertises its garments by promoting the benefit of several layers of clothing (Figure 4.13). In their garments, they emphasise the following features:
- the 'wicking' properties of the inner layer (next to the skin);
- the choice of several different layers depending on external conditions;
- the importance of a windproof and/or waterproof outer layer;
- the overall light weight of their garments.

Imagine you are walking for several hours in the mountains, in wet, windy conditions with temperatures near to freezing, and you are wearing this type of clothing.

1 What do you think the manufacturer means by 'wicking' and why is this beneficial?
2 How do several layers keep you warm?
3 Why is it important that the outer layer is windproof?
4 Explain what features should be incorporated into the design of the waterproof material to make sure it is effective in these conditions.
5 Explain why it is important to pay attention also to the design of hats, gloves and boots worn in these conditions.
6 Why are lightweight garments preferable to heavy clothing?
7 How far do you think the manufacturer's claims about their garments are biologically sound?

A brief review – responses to changes in temperature

Let's review the chapter so far and make a brief summary of how the body responds to changes in temperature, including extremes of heat and cold in the surrounding environment. The essentials are listed below.

- Exchange of heat between the body and surrounding environment occurs by conduction, convection, radiation and loss of heat through evaporation of sweat.
- The hypothalamus helps to control the body temperature through physiological mechanisms.
- Heat gain within the body is derived mainly from metabolic activity, including muscular activity and shivering.
- Reduction of heat loss (when it is cold) is mainly through vasoconstriction of blood vessels (arterioles) in the skin, so that less blood flows near the skin surface and less heat is lost from the skin.
- Increase in heat loss (when it is hot) is mainly through vasodilation of blood vessels (arterioles) in the skin, so that more heat is lost from the skin (compare with vasocontriction, above).
- In hot conditions, there is an increase in sweat production and this becomes the most important mechanism for cooling the body and maintaining a constant body temperature.
- As people acclimatise to hot conditions, there is evidence that the rate of sweating is higher and that the sweat produced has a lower salt concentration.
- In extremes of both hot and cold, behavioural mechanisms are very important in enabling people to live normally and to survive in exceptional conditions outside that experienced in their usual living places. Behavioural mechanisms include patterns of activity (including physical exercise), clothing and the artificial environment created inside buildings (see Question box on this page).
- When the body can no longer cope with temperature extremes, it suffers from stress.
- Under hot conditions, the symptoms of heat stress arise mainly from the degree of dehydration suffered by the body together with the loss of salts, until the point is reached when there is a breakdown of temperature regulation mechanisms. The progression is through heat collapse to heat exhaustion and finally to heat stroke.
- Under cold conditions, cold stress can be apparent as actual damage to body tissues (described as cold injury) and as a fall in body temperature beyond acceptable limits, known as hypothermia.
- Recovery is possible from extremes of both heat stress and cold stress, but death can result in severe cases or if appropriate treatment is not undertaken.

QUESTION

At the start of this chapter, we referred to the fact that many animal species have a rather narrow range of tolerance for their natural habitat, whereas the human species inhabits places with much wider ranges of temperatures.

- What evidence is there that the body adjusts (becomes acclimatised) through *physiological* mechanisms in extremes of heat and cold?
- List behavioural mechanisms shown by people in response to extremes of heat and to extremes of cold. In your list, include reference to patterns of activity, clothing and buildings.
- How far do you think this wider distribution of humans is a result of development of *behavioural* mechanisms that enable them to cope with these extremes of temperature?

High altitude

BACKGROUND MATERIAL

Climbing to the top of the world

Mount Everest, or *Chomolungma* ('Goddess Mother of the Earth') in Tibetan, was first climbed in 1953. The climbers were part of a large-scale expedition, with many people involved in the team. The whole expedition required considerable planning and meticulous attention to detail with regard to routes to be taken, specialised equipment required, food stores and oxygen supplies. When Hillary and Tenzing stood at 8848 m, on the summit of the highest peak in the world, their success showed that humans can overcome the extremes of physical conditions associated with the highest mountains, in terms of exposure to cold, low levels of oxygen and dangerously hazardous terrain (Figure 4.14).

Since then, a relatively small number of people have climbed Mt Everest solo and without the help of extra oxygen: two who have achieved this are Reinhold Messner in 1980 and Alison Hargreaves in 1995. An increasing number of other climbers have now reached the summits of Everest and of the remaining peaks over 8000 m. Many, however, have endured the hardships without the success. Some have lost their lives in the attempt (including Alison Hargreaves a few weeks after her success on Everest); others have become victims of frostbite and sacrificed toes or fingers. Nobody can stay for long at those heights: realistically a few hours at the most. The urgent need is to summon both the physical energy and mental concentration to make a safe descent to lower altitudes.

By the time of the 50th anniversary (29 May 2003) of the first ascent, more than 1200 people had reached the peak of Mt Everest. Over 175 have died in the attempt. The records now go to the most rapid ascent. From Base Camp at 5400 m to the peak at 8848 m, a time of just 12 hours 45 minutes has been logged, but this is sure to be surpassed. Compared with the slow progress of the large-scale early expeditions, the feat is remarkable – the height of the summit and the effects of high altitude do not change!

Figure 4.14 Mountaineers face difficult terrain and are exposed to harsh physical conditions

People at high altitude

People **native** to high altitudes are seen as rugged, robust and sturdy, able to live successfully and in equilibrium with the particular conditions characteristic of high altitude. These qualities may be due partly to the outdoor lifestyle associated with the mountain environment, but also indicate that the natives are well adapted to the rigours of life at high altitude. For convenience, we will define 'high altitude' as being above 3000 m. Using this as an arbitrary limit, we find the main native high-altitude populations living in the Himalayan region of Asia (notably Tibetans and Nepalese) and in the Peruvian Andes of South America (Quechua Indians) (Figure 4.15). The upper limit for permanent settlements is about 4500 m in both regions, though to some extent this is a reflection of historical events and of accessibility and not determined solely by the altitude itself.

Figure 4.15 Natives in high mountain regions: (left) shepherd in Himalayan region of north-west India (4000 m); (centre) Tibetan woman, western China (3200 m); (right) Quechua Indian woman and child, Peru (4000 m)

Visitors to these high altitudes from the lowlands may suffer varying degrees of discomfort and illness (Figure 4.16). With increasing numbers of tourists visiting high mountain regions for short periods, these effects are becoming more familiar. If, on a quick trip up Mount Kenya, you spend a night below the peaks in a hut at about 4500 m, you are likely to have a persistent headache, probably slight nausea and an extreme disinclination to continue with your journey the next day. Trekkers in the Himalayan mountains, from altitudes of about 3000 m and higher, find that their movements become slower. They are likely to stop frequently and also experience difficulty in maintaining steady breathing, or may become very conscious of it. These symptoms are known as **mountain sickness** (see pages 80 and 81 and Figures 4.16 and 4.23). Similar difficulties have been encountered by military personnel unaccustomed to life at high altitude, for example, when controlling border disputes in the Himalayan region. The problems are worst during the first few days at high altitude and can be much more severe if the ascent is rapid, perhaps by motor vehicle or even by plane. Bus passengers in the Andes are sometimes supplied with oxygen when their journey takes them over high passes. Certainly, the best way to get into the mountains is a slow plod on foot. The effects usually disappear rapidly on descent to lower altitudes.

Environmental conditions at high altitude

The physical conditions in high mountains that lead to the symptoms of mountain sickness are dominated by low levels of oxygen, known as **hypoxia**. In addition, the body has to withstand low temperature, high winds, low humidity and increased solar radiation. These physical conditions impose considerable stress on the human body.

Figure 4.16 Visitors on Mount Kenya at about 4000 m already struggling and losing interest with still another 1000 m of height to climb to reach Point Lenana, one of the summits

Low atmospheric pressure

Atmospheric pressure decreases with increasing altitude. This means that the air at sea level contains more molecules per unit volume than at higher altitudes, up to the peaks of the highest mountains and beyond. Oxygen is one of the gases in air, so the oxygen molecules in the air contribute to atmospheric pressure. The part of the pressure due to oxygen is called the **partial pressure of oxygen** and this is written as **pO_2**. For humans, the 'partial pressure of oxygen' determines how much oxygen is available at the lung surface for uptake and transport into the body tissues.

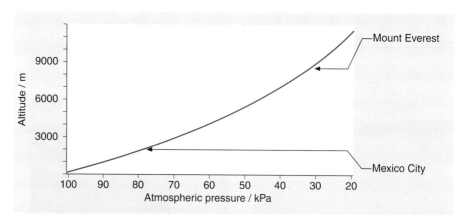

Figure 4.17 Changes in air pressure with altitude

At sea level, atmospheric pressure is approximately 100 kPa and Figure 4.17 shows how it becomes less at higher altitudes. The proportion of oxygen in the air is approximately 21 per cent, so we can use data in Figure 4.17 to calculate values for pO_2 (partial pressure of oxygen) at different altitudes. Table 4.1 shows this for three altitudes: sea level, 3500 m and at 8500 m. You can see that the value for the partial pressure of oxygen at 8500 m is only about one-third of its value at sea level. This means that, near the summit of Mt Everest, very much less oxygen is available for gas exchange at the lung surface.

Table 4.1 *Changes in atmospheric pressure and partial pressure of oxygen at different altitudes. Note that 3500 m is about the height at which those living permanently are considered to represent 'high altitude people' and 8500 m is approaching the height of Mt Everest. To obtain the data in the table, you can read the figure for atmospheric pressure from the graph (Figure 4.17) then calculate 21 per cent of this to give the value for the partial pressure of oxygen (pO_2)*

Height / m	Atmospheric pressure / kPa	Partial pressure of oxygen / kPa
sea level	100	21
3500	65.5	13.8
8500	31.9	6.7

Low temperature

Air **temperature** drops approximately 1 °C for each 150 m of ascent, though this varies in different mountain ranges around the world. However, a rough estimate indicates that, compared with sea level in a nearby area or equivalent latitude, the temperature at 3500 m is likely to be 23 °C lower, and at 8500 m 56 °C lower. This emphasises the need, at very high altitudes, to conserve body heat by providing good insulation in terms of clothing worn and housing, or portable tents in the case of mountaineers.

Low humidity and high winds

At high altitudes, **humidity** of the air is often low. This leads to an increased loss of heat from the body as a consequence of evaporation of sweat. Mountaineers frequently suffer from dryness and cracking of the lips or other exposed parts of the skin. There is also a tendency to develop a persistent cough, because breathing is usually through the mouth and the air is cold and dry. Loss of water from the body by evaporation brings a danger of dehydration. In addition, strong winds are frequently associated with mountain regions, which adds to the potential heat loss from the body due to the wind chill factor.

Solar radiation

At high altitudes there are fewer molecules (of oxygen, nitrogen and ozone) per unit volume of the atmosphere and this leads to an increase in solar radiation reaching the ground. These gas molecules effectively absorb and scatter radiation at various wavelengths. Ultraviolet radiation at 3000 m is about double that at sea level because of the reduced ozone at high altitude. Excessive ultraviolet radiation may damage the cornea of the eye, a condition known as **snow blindness**, and mountaineers frequently wear dark goggles to protect their eyes. Exposure to ultraviolet radiation may lead to a higher incidence of skin cancer; reflection from snow intensifies the effects of radiation on exposed skin. During the daytime the high solar radiation can provide additional heat for the body, though temperatures fall rapidly at night. A climber on Mt Everest at 8530 m was able to remove his down-filled clothing without suffering from cold when in the full sun and at low wind velocity.

Physiological effects of high altitude

Hypoxia

Because of the low partial pressure of oxygen in the atmosphere at high altitudes, less oxygen is available for the body and less reaches the tissues. These circumstances lead to the condition known as **hypoxia** in the body. Obtaining enough oxygen for respiration is of crucial importance in these situations. The body's responses demonstrate different ways of overcoming the effects of hypoxia. These include:

- hyperventilation
- increased pulmonary diffusing capacity
- increased transport of oxygen in the blood
- changes in cardiac output.

The term **hypoxia** is used to describe a low partial pressure of oxygen, which is insufficient for haemoglobin to become fully saturated. This may be due to a low partial pressure of oxygen in the air, for example, at high altitude where, although the percentage of oxygen in the air is normal, the air pressure is low.

Oxygen is used in cells in the process of respiration, which releases energy for the many metabolic and physical activities of the body. Any reduction in available oxygen leads to impairment of a range of body functions, and, in extreme cases, to death. To reach the mitochondria in the cells where the respiratory reactions take place, the oxygen molecules must pass through a series of barriers. There is effectively an oxygen **cascade**, down a gradient of pressure from the atmosphere to the inside of the cells. The stages for transfer of oxygen are summarised in Figure 4.18.

Figure 4.18 The cascade of oxygen from alveoli to body cells: summary of respiratory processes

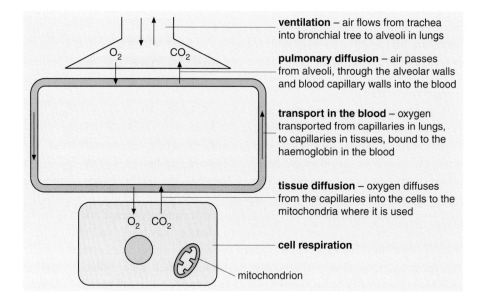

ventilation – air flows from trachea into bronchial tree to alveoli in lungs

pulmonary diffusion – air passes from alveoli, through the alveolar walls and blood capillary walls into the blood

transport in the blood – oxygen transported from capillaries in lungs, to capillaries in tissues, bound to the haemoglobin in the blood

tissue diffusion – oxygen diffuses from the capillaries into the cells to the mitochondria where it is used

cell respiration

mitochondrion

The partial pressure of oxygen at each stage is a measure of its availability. For a healthy person at sea level, the difference between the partial pressure (pO_2) of inspired air and the final pO_2 in the blood capillaries surrounding the body cells is sufficiently great for there to be no difficulty for the oxygen to reach the mitochondria. However, at higher altitudes, because of the lower pO_2 of oxygen in the atmosphere, this gradient is reduced considerably. The body responds in several ways to overcome or compensate for the effects. Native high-altitude people have already adjusted to the lower pO_2 of oxygen, whereas visitors beome **acclimatised** and usually adjust after a number of days.

The gradient in pO_2 at these different stages is shown in Figure 4.19. It compares the fall in pO_2 for a person living at sea level with that of a native highlander living at 4540 m and a 'visiting' climber at 6000 m. When air enters the bronchial tree inside the lungs, it becomes saturated with water vapour, which itself exerts a partial pressure. Expired carbon dioxide in the lungs also exerts a partial pressure and these two gases account for the immediate steep fall in pO_2 inside the alveoli.

Haemoglobin in the red blood cells combines with oxygen to form oxyhaemoglobin. The reaction between haemoglobin and oxygen is summarised in the equation

$$\underset{\text{haemoglobin}}{\text{Hb}} \;+\; 4O_2 \;\rightleftharpoons\; \underset{\text{oxyhaemoglobin}}{\text{HbO}_8}$$

From the dissociation curve shown in Figure 4.20, we can see that at high partial pressures of oxygen the haemoglobin takes up oxygen, but at low

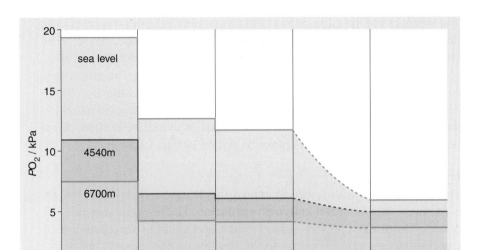

BACKGROUND MATERIAL

A reminder about dissociation curves

(See also Chapter 2, page 47)

The term **percentage saturation** is used to give a measure of the amount of oxyhaemoglobin in relation to (deoxy)haemoglobin in the blood. In practice, saturation of haemoglobin with oxygen is rarely higher than about 95 per cent. The percentage saturation varies with the partial pressure of oxygen but is not directly proportional to it. Expressed as a graph, we might expect a straight line, but the typical sigmoid shape of such curves (known as dissociation curves) is due to the way in which the affinity of haemoglobin for oxygen alters after the binding of the first oxygen molecule.

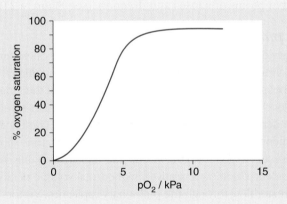

Figure 4.20 Oxygen dissociation curve of haemoglobin showing the percentage saturation with increasing partial pressure of oxygen, at 37 °C and pH 7.4

partial pressures oxygen is unloaded or released from the haemoglobin. Thus oxygen is collected by haemoglobin in the blood flowing through the capillaries in the lungs at the alveolar surface where partial pressure are high. Oxygen is then off-loaded at low partial pressure in the tissues at the surface of the cell. From there it passes to the mitochondria where it is used in respiratory reactions.

The steepness of the curve at low partial pressures indicates that a very small drop in partial pressure allows large amounts of oxygen to be released. The critical partial pressure of oxygen needed at the mitochondria for oxidative reactions to take place is less than 0.4 kPa. If we look at the flat part of the curve, we can also see that the partial pressure of oxygen in the alveoli can drop from 13.3 to 8 kPa without making any appreciable difference to the degree of saturation of the haemoglobin. Thus, at moderate altitude, people do not suffer from a reduced uptake of oxygen by haemoglobin.

Hyperventilation

Visitors to high altitudes notice an immediate increase in their rate of breathing, even at rest. This is known as **hyperventilation**. The response may be observed from about 3000 m upwards and is also found in native highlanders. Quechua Indians in the Peruvian Andes breathe at a rate about 30 per cent higher than people at sea level, though visitors maintain a rate higher than that of the native highlanders. As well as more rapid breathing, the breaths may be deeper. This hyperventilation increases the ventilation inside the alveoli and increases the pO_2 of oxygen at the boundary between the alveoli and the blood.

Another consequence of hyperventilation is increased exhalation of carbon dioxide. This results in a degree of **alkalaemia**, or abnormally high pH in the blood. (Carbon dioxide dissolves in the blood to form carbonic acid, which increases the acidity, or lowers the pH.) The acid–alkali balance in the blood must be restored to avoid further physiological effects. A complication arises because the level of carbon dioxide in the arterial blood is one factor that controls the rate of breathing (hence ventilation). As the carbon dioxide level in the blood falls, the ventilation rate decreases until a high carbon dioxide level builds up and the rate increases again. This may lead to irregularity in breathing. The low carbon dioxide level in the blood (or high blood pH) triggers the symptoms of nausea and dizziness (see Figure 4.23).

Increased pulmonary diffusing capacity

The pulmonary diffusing capacity is a measure of the rate of exchange of gases between the alveoli and the (pulmonary) capillaries surrounding them. The capacity can be improved by an increase in surface area of the lung and by an increase in blood flow in the surrounding capillaries. There is some evidence that native highlanders have a lung volume larger than that of comparable lowlanders. This enlargement gives an increased alveolar surface area, thus allowing a higher rate of diffusion of oxygen from the alveoli into the surrounding blood capillaries. Native highlanders also show a relatively higher total volume of blood, which results in a greater flow through the pulmonary capillaries. Visitors to high altitudes do not seem to develop a similar increase in lung capacity; if anything there at first appears to be a decrease in vital capacity (maximum volume that can be exhaled), though this change is reversed after a few weeks at high altitude.

Increased transport of oxygen in the blood

The total quantity of oxygen transported in the blood to the cells depends on three factors:

- the cardiac output
- the haemoglobin concentration
- the saturation of the haemoglobin with oxygen.

Cardiac output (Figure 4.21) may be altered by changes in both the **heart rate** (number of heart beats in a given time) and the **stroke volume** (volume of blood pumped out at each beat). Increases in one or both of these attributes result in more blood being pumped through the pulmonary capillaries, allowing an increase in the collection of oxygen from the alveoli and increased delivery to the cells. Visitors to high altitudes show an immediate increase in heart rate, though stroke volume remains steady at first then appears to fall. Overall cardiac output thus increases for the first few days then returns to about the same as that at sea level. In native highlanders the resting cardiac output is about the same as that for comparable lowlanders.

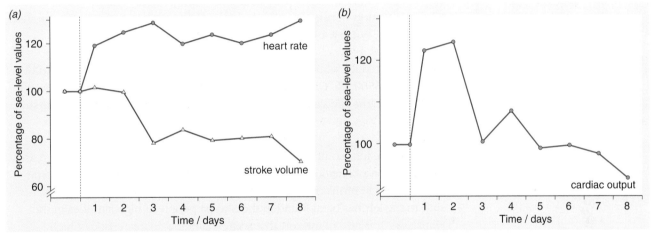

Figure 4.21 Changes in cardiac output of a group of people during the first 8 days after exposure to an altitude of 3800 m (cardiac output = heart rate × stroke volume): (a) heart rate and stroke volume; (b) cardiac output

Native highlanders, compared with equivalent lowlanders, show higher levels of both the haemoglobin concentration in the blood and the total number of red blood cells (Table 4.2). A similar increase is noted when lowlanders visit high altitudes. The response is due to increased production of red blood cells,

Table 4.2 *Red blood cell (rbc) counts and haemoglobin (Hb) concentration at different altitudes (all values are for adult males)*

Altitude	Rbc count / dm⁻³	Hb concentration / g dm⁻³
Sea level (normal values)	5.0×10^{12}	148
4540 m (natives in the Andes)	6.4×10^{12}	210
5790 m (mountaineers on Mt Everest)	5.6×10^{12}	196

which takes place in the bone marrow, stimulated by the hormone erythropoietin. Increased secretion of erythropoietin occurs in conditions of low oxygen. There is also an overall increase in blood volume. At 4500 m, these values are about 25 to 30 per cent above the corresponding values at sea level. In visitors, the higher red blood cell count is noticeable about 3 to 4 days after arrival at high altitude, though the increase in haemoglobin and blood volume continues for several weeks. With more haemoglobin in the blood, the capacity for carrying oxygen is increased, provided the pO_2 of oxygen is high enough for the haemoglobin to collect the oxygen (become saturated). The increased blood volume noticed at higher altitudes is due to the increased red blood cell volume. Plasma volume actually decreases. A potential danger of the raised red blood cell count is that the blood becomes more viscous (thicker), which in extreme situations may reduce the flow of blood through the capillaries.

A closer look at the oxygen–haemoglobin dissociation curves reveals how readily the haemoglobin collects oxygen and how easily it gives it up again to the cells. If the curve lies further to the right, oxygen is released more easily, but the percentage saturation of haemoglobin is lower, so the haemoglobin actually carries less oxygen. If, on the other hand, the curve lies further to the left, the reverse situation occurs – the haemoglobin has a higher percentage saturation with oxygen, but releases the oxygen less easily (see Figure 4.22).

A shift to the right would provide some advantage at moderate high altitude (3000 m to 5500 m). Quechua Indians living in the Peruvian Andes do show a shift to the right and visitors to high altitudes are likely to show a similar shift to the right.

A shift to the left would be favoured at very high altitudes because it increases the saturation of haemoglobin with oxygen, though there is a corresponding disadvantage in terms of the point at which oxygen is released to the cells. A shift to the left has been found in the Sherpa people at high altitudes in the Himalayas. It is also of interest that some mammals characteristic of high altitudes, such as the llama, show a shift to the left of the oxygen–haemoglobin dissociation curve. A shift to the left suggests adaptation whereas a shift to the right indicates acclimatisation.

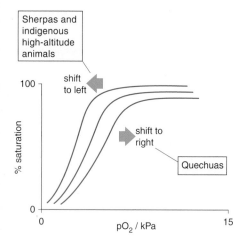

Figure 4.22 Oxygen dissociation curve for high-altitude inhabitants (Quechuas and Sherpas). A shift to the left suggests adaptation; a shift to the right indicates acclimatisation. Visitors to high altitude are likely to show a shift to the right.

High altitude stress

As described earlier in this chapter, native highlanders and visitors to high altitude are exposed to harsh environmental conditions and this exposure imposes stress on the body. The dominating stress comes from low levels of oxygen. The effects of and responses to hypoxia are described in the section above. Here we look at other effects of high altitude stress under the following headings:

- general symptoms of mountain sickness
- redistribution of body fluids and secretion of ADH (antidiuretic hormone)
- mental reactions.

Mountain sickness

These symptoms described for visitors to high altitudes (see page 73) can be attributed primarily to the lower levels of oxygen, though some effects are a result of unfamiliar physical demands of mountain terrain and, at extreme heights, of the cold. The symptoms, collectively known as **acute mountain sickness**, include: headaches (mild to severe); lack of concentration and giddiness; coughing and difficulty with breathing; palpitation and very rapid beating of the heart; loss of appetite, feelings of nausea and vomiting; muscular weakness, exhaustion and poor coordination; reduced production of urine, swelling, particularly of the legs, and also development of excessive fluid in the lungs; a general disinclination for exertion or coherent thinking and frequent difficulty in sleeping (Figure 4.23). At extreme heights, or in severe cases, loss of consciousness or death may occur, unless the sufferer is brought rapidly to a lower level.

Development of symptoms of mountain sickness depends on the rate at which the ascent is made but the severity is unpredictable, and fit young males are often more prone to suffering than the middle-aged or females. It must be emphasised that the best (and sometimes the only) way to recover from mountain sickness, is to descend to a lower altitude, and to do so quickly.

Redistribution of body fluids

Distribution of fluids in the body is influenced by a number of interacting factors, including blood pressure, blood volume, salt balance and hormones. The changes are complex and some are affected by hypoxia, as experienced at high altitude.

One of these hormones is antidiuretic hormone (ADH), secreted by the posterior pituitary gland. ADH increases the reabsorption of water by the kidneys. This results in more water being retained in the body and a smaller volume of urine being produced. In **mild hypoxia**, there is probably a decrease in secretion of ADH, which results in more urine being produced, a situation known as **diuresis**. However, in conditions of **severe hypoxia**, increased secretion of ADH has the opposite effect, leading to retention of water in the blood and reduced urine production. At high altitude there is a redistribution of the circulation of the blood, with a reduced flow to the extremities.

The excess fluid retained in the body tends to accumulate outside the blood vessels, particularly in the lungs and brain. The condition is known as **oedema**. In severe **pulmonary oedema**, fluid collects in the lungs and the sufferer may froth at the mouth and become very breathless. In **cerebral oedema**, the brain swells with fluid and then presses against the cranium (the part of the skull which contains the brain). This induces severe headaches, leading to loss of consciousness and sometimes to death. For recovery, in both situations, it is essential for the person to be taken rapidly to a lower altitude. Milder forms of oedema show as a puffiness in the face, particularly around the eyes, and as swelling in the legs and feet.

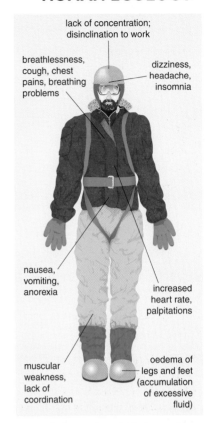

Figure 4.23 Mountain sickness and its symptoms

Mental reactions at high altitude

Visitors to moderately high altitudes frequently show slower mental reactions and weaker decision-making ability than would be expected from the same people at lower altitudes. This can be tested by a variety of memory tests. There are many stories of mountaineers making foolish mistakes at high altitudes, which may often have been the cause of serious accidents in this difficult terrain. These mental reactions may have been influenced by a combination of the direct effect of hypoxia on brain function as well as stress imposed on the body by the onset of symptoms of mountain sickness. At altitudes over about 5500 m, the hypoxia has yet more severe effects on the activity of the brain, leading to loss of concentration and inability to carry out calculations or make reliable judgements. At still higher altitudes, hallucinations may become apparent, often manifesting themselves as the presence of another companion on the mountain. Habeler and Messner, when they climbed Mt Everest without oxygen in 1979, experienced a feeling of euphoria, somewhat dangerous at a height of 8848 m, surrounded by snow, ice and precipitous slopes. To quote from Habeler, 'I felt somehow light and relaxed, and believed nothing could happen to me. Undoubtedly, many of the men who have disappeared for ever in the summit region of Everest have also fallen victim to this treacherous euphoria'.

Acclimatisation or adaptation?

Native highlanders, such as the Quechua Indians in the Andes or Tibetans and Sherpas in the Himalayas, are able to lead active and physically demanding lives at altitudes in excess of 3000 m. Newly arrived lowland visitors to high altitudes notice an immediate decline in their normal physical activity, and though their performance gradually improves, it may never reach that of the native highlander, or at least not for a considerable time. Athletes competing in the 1968 Olympic Games in Mexico City (2380 m) experienced difficulties in reaching their maximum performance. However, athletes who have trained at moderate or high altitudes, or who are normally resident there, may show superior performance (at low or at high altitudes), particularly in middle-distance or long-distance running events. Sherpas are known for their support as porters in mountaineering expeditions: at about 3000 m they can carry loads of about 60 kg and even at 7000 m can manage loads of about 18 kg. This is more than would be expected from a newly arrived lowlander and is certainly related to the Sherpa's superior capacity to utilise oxygen.

If native highlanders were **genetically adapted** to high altitude, the features that allow normal activity in the hypoxic conditions at high altitudes would be inherited and irreversible. There is little evidence to support this, nor is there evidence that the features would persist if the highlander went to live at sea level. It is preferable to consider both the natives and the visitors as showing different degrees of **acclimatisation** to the high altitudes. Acclimatisation implies that reversible non-inherited changes are shown in structural and physiological features in the body, in response to the conditions experienced at high altitude, and these changes help survival. Thus the features which allow success in hypoxic conditions are of the same kind in both native highlanders and lowland visitors. However, in the native highlander these features probably developed during normal growth and in the early years of childhood, and so appear to be permanent, whereas in the visitor the features are acquired over a period of time after arrival at high altitude.

A few examples (discussed earlier in this chapter) will help to illustrate this.
- Native highlanders show an increased lung volume which contributes to a higher pulmonary diffusing capacity. This would develop in childhood, though perhaps the feature is partly a response to the physical exercise associated with the lifestyle in mountainous regions. A lowland visitor is unlikely to show an increase in lung volume.
- A 'barrel-shaped' chest with large lung volume is a characteristic of Quechua Indians. (This feature may have already existed and given this group an advantage when they first colonised the high regions in the Andes.)
- Both highlanders and visitors hyperventilate, though the visitor maintains a higher rate for a longer period than the native highlander.
- At high altitudes both groups show a higher red blood cell count and higher haemoglobin concentration compared with similar groups living at sea level.

On descent to sea level, several of the features apparent at high altitude, including hyperventilation, haemoglobin level and cardiac output, fall to a normal level. Native highlanders and visitors to high altitudes are affected in a similar way.

Certain species of mammals, including yaks and llamas, live permanently at high altitudes, so it may help to look at their features in comparison to native highlanders to understand the extent to which each is adapted or acclimatised. Comparing llamas with Quechua Indians at similar altitudes, llamas do not hyperventilate, nor do they show as much increase in haemoglobin concentration. However, llamas are efficient in their extraction of oxygen by the tissues at low partial pressures and show a shift to the left of the oxygen dissociation curves. Sherpas show a similar shift, but the Quechua Indians do not. The llama is considered to be adapted, the Sherpas are on their way to being fully adapted rather than acclimatised, and the Quechua Indians can best be described as acclimatised. The degree of adaptation is relative and related to the length of time the group has lived at high altitude. The lowland visitor is always at a disadvantage during the first few days at high altitude, but becomes at least partially acclimatised after a period of time.

A brief review – response to high altitude

Let's review the high altitude section of this chapter and make a brief summary of how the body responds to life at high altitude, giving consideration both to natives who live permanently at high altitudes and to visitors who come for shorter periods of time. The essentials are listed below. Comparisons with people living at sea level are implied in the various statements.

- We use 3000 m as an arbitrary limit and describe people living above this height as high altitude people. The upper limit for permanent settlements is about 4500 m. The highest point reached by climbers is 8848 m, the summit of Mt Everest.
- Environmental conditions experienced at high altitudes include low atmospheric pressure, low temperatures, low humidity, high winds and increased solar radiation. These physical conditions can impose stress on the human body.

Figure 4.24 Domesticated llamas in Bolivia at over 4000 m, where they are used by the Aymara people for meat, wool and to some extent as pack animals. Like the Quechua Indians, the Aymaras are native high altitude people and are characterised by huge chests and lungs, which enable them to live successfully at high altitudes.

- Physiological problems that may result from exposure to low temperatures, high winds and low humidity are similar to those described earlier in the chapter, under extremes of temperature. The need for suitable clothing to provide protection becomes a vital necessity at very high altitudes.

- Low atmospheric pressure is the main factor that limits life and activity at high altitudes. Because of the lower partial pressure of oxygen (pO_2), less oxygen is available for exchange at the surface of the lung.

- The body responds to this lower level of oxygen in a number of ways to enable enough oxygen to reach the cells of the body.

- At high altitudes, a person is likely to breathe at a faster rate and with deeper breaths (hyperventilation).

- High altitude natives may show a higher rate of exchange of gases between the alveoli and the surrounding capillaries (increased pulmonary diffusing capacity).

- The heart beats at a faster rate and pumps out more blood with each beat. This increase in cardiac output means more blood is pumped out of the heart, carrying oxygen from the lungs to the body cells.

- Natives and visitors at high altitudes usually have relatively higher numbers of red blood cells and a higher concentration of haemoglobin.

- Oxygen dissociation curves show how much oxygen is taken up and how easily it is released at the surface of the cells. A shift to the right indicates that oxygen is released more easily whereas a shift to the left indicates a higher percentage saturation of haemoglobin with oxygen.

- A shift to the right is associated with people showing acclimatisation whereas a shift to the left is associated with people (and some high altitude mammals) showing adaptation.

- Visitors acclimatise to some extent after a few days at high altitude. The effects wear off on return to lower altitudes. Natives show responses similar to those of visitors, and in some parts of the world appear to show permanent adaptations to high altitude conditions. Examples of features shown by natives are listed on pages 82 and 83.

- As a person ascends to higher altitudes, the body suffers increasingly from stress.

- Symptoms of high altitude stress are known as mountain sickness – or acute mountain sickness in severe situations.

- The various symptoms of mountain sickness are summarised in Figure 4.23 (on page 81). They include headaches, nausea, dizziness, breathlessness, palpitations and oedema. They are caused mainly by the low level of oxygen (known as hypoxia) and the effect on distribution of fluids in the body.

- Mental reactions are slower at high altitudes.

- The best way to recover from mountain sickness is to descend to lower altitudes as fast as possible. With acute mountain sickness, a person may become unconscious and death can result in severe cases or if appropriate treatment is not undertaken.

QUESTIONS

Turning your theory into practice – some practical questions!

Some university students studying biology are planning an expedition in the Himalayan mountains. They will do some botanical work, surveying distribution of plants and studying their ecology, and also aim to climb some of the peaks in the area. They expect to spend about 6 weeks at altitudes between 2000 m and 6000 m.

Write about a page of 'advice', which can be circulated to the students intending to join the expedition. The advice should include:
- an indication of the conditions expected as they reach the higher altitudes
- how the body is likely to react at the higher altitudes and over the period of time spent there
- a biological explanation of the changes taking place in response to the higher altitudes
- how best to prepare for the journey and how to take care of themselves during the expedition.

Don't forget it can be cold at high altitudes, so include reference to suitable clothing.

Then list some simple measurements they could take or observations they could make during the expedition to monitor the responses of the body to higher altitudes.

Remember that the students will be carrying most of their own baggage so will not be able to take heavy or complex equipment with them, but you can include some measurements that could be taken before they depart and again on their return.

Sexual reproduction

The production of new individuals of the same species is a fundamental characteristic of living organisms. Reproduction allows for the replacement of individuals that die and ensures the continuity of the species. If conditions are favourable, it can also result in an increase in numbers. Genetic information from the individuals of one generation, the **parental generation**, is passed on to the next generation, the **offspring**, ensuring that the characteristic features of a species are perpetuated.

Reproduction may be asexual or sexual. **Asexual reproduction** usually involves:
- a single individual
- no gamete formation
- the inheritance of identical genetical information from the parent
- mitotic divisions.

The offspring may be referred to as **clones**.

The only genetic variation possible arises from random mutations. Few animal species reproduce naturally in this way, although both animals and plants have been cloned successfully using artificial techniques.

Sexual reproduction involves:
- the formation of **gametes** by individuals of the same species
- a type of division called **meiosis** which halves the number of chromosomes
- the fusion of a **haploid** male gamete with a haploid female gamete to form a **diploid zygote**
- the growth of the zygote by mitotic divisions into a new individual.

> **DEFINITION**
>
> A **haploid** cell has only one set of chromosomes. A **diploid** cell has two sets of chromosomes.

Halving the chromosome number during gamete formation ensures that new individuals have the same number of chromosomes as the parents. The offspring produced by sexual reproduction show genetic variation. During meiosis, there may be some exchange of portions of chromosomes, as they separate from each other, leading to slight differences in the genes. This results in each gamete being unique. Random fusion of male and female gametes also introduces genetic variation.

Meiosis

Before considering the stages in the process of meiosis in detail, it is helpful to understand the nature and origin of the chromosomes in a diploid cell. In a diploid organism, one set of chromosomes will have come from the male parent and the other set from the female parent. There will be two copies of each chromosome in the cells. These chromosomes are referred to as **homologous**.

Two homologous chromosomes share the following characteristics:
- they are exactly the same length
- they have the centromere in the same position
- they contain the same number of genes
- the genes are arranged in the same linear order.

When mitosis occurs, the homologous chromosomes act independently. However, in meiosis, the homologous chromosomes pair up during the first stage. Their subsequent separation ensures that one of each pair is present in the gametes that are formed. Figure 5.1 illustrates the roles played by mitosis and meiosis in a life cycle.

Meiosis differs from mitosis in that it involves a reduction in the number of chromosomes from the diploid number (2n) to the haploid number (n). Replication of the DNA takes place in interphase, but this is followed by two cycles of nuclear division. One cycle separates the homologous chromosomes and the other separates the chromatids. These are usually referred to as the first and second meiotic divisions and result in the formation of four haploid nuclei from one diploid nucleus. This type of division occurs during **gametogenesis**, the production of sperm and ova in animals, and at some stage in the formation of pollen grains and ovules in plants. As mentioned earlier, the gametes will have a single set of chromosomes, but these will not be identical. During the initial stages of meiosis, when the homologous chromosomes pair up, there is the possibility of exchange of sections of non-sister chromatids, leading to variation.

The events that take place during meiosis form a continuous process, but as with mitosis, they are separated into different stages for ease of description. The same stages occur, but as there are two divisions of the nucleus, each stage occurs twice (see Practical: Meiosis in pollen mother cells). For example, prophase I refers to the first meiotic division and prophase II to the second meiotic division.

The stages of meiosis
Prophase I
In prophase I, the chromosomes condense and become visible as single threads. It is not possible to distinguish that each chromosome consists of two sister chromatids in the early stages. Each chromosome appears to have a beaded appearance due to localised coiling of the DNA. Homologous chromosomes pair up precisely along their length, a process known as **synapsis**. More condensation makes them appear shorter and fatter, and it is now possible to see that each chromosome consists of two chromatids. Each pair of homologous chromosomes is called a **bivalent**.

As the chromatids of each pair become visible, the members of a pair seem to repel each other in some regions but remain attracted at others (Figures 5.2 to 5.4). At the points of attraction, non-sister chromatids break and rejoin at exactly corresponding locations. This is known as **crossing over** and results in the formation of **chiasmata** (sing. **chiasma,** from the Greek for 'cross-arrangement').

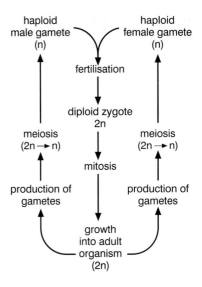

Figure 5.1 The significance of mitosis and meiosis in a life cycle

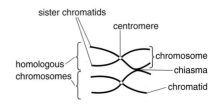

Figure 5.2 A pair of homologous chromosomes (a bivalent) during meiosis

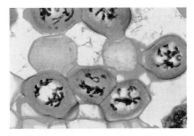

Figure 5.3 Photomicrograph of prophase I of meiosis in Lilium, *showing the chromosomes as threads, with some regions of contact between chromatids*

(a) **Early prophase** in an animal cell showing the pairing of homologous chromosomes

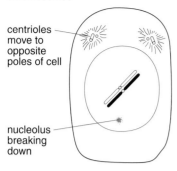

centrioles move to opposite poles of cell

nucleolus breaking down

(b) **Late prophase** – spindle organisation has begun

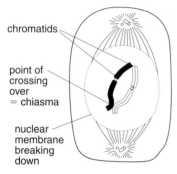

chromatids

point of crossing over = chiasma

nuclear membrane breaking down

Figure 5.4 Diagrammatic representation of prophase I of meiosis, showing that the regions of contact between chromatids (chiasmata) are points where genetic material crosses over when the chromatids break and rejoin (only one pair of chromosomes is shown)

More than one chiasma can form and it may involve the same pair or another pair of non-sister chromatids. If one chiasma forms, the bivalents appear in the shape of a cross, if two, a ring shape and if more than two, a series of loops arise at right angles to each other. Chiasmata hold the two homologous chromosomes together. During the later stages of prophase I, the nucleoli disappear, the centrioles migrate to opposite poles of the cell and the nuclear envelope breaks down. Spindle formation begins as microtubules become arranged across the middle of the cell from pole to pole.

Metaphase I

The bivalents become attached to the spindle in metaphase I (Figures 5.5a, 5.6a). The centromeres attach to individual spindle fibres and the bivalents are moved so that they are arranged along the equator of the spindle. They are orientated so that the centromere of one homologous chromosome will move to one pole and the other will move to the opposite pole. At this point, the centromeres of each homologous pair repel each other strongly, but the sister chromatids are closely associated.

Anaphase I

In anaphase I, the homologous chromosomes are separated (Figures 5.5b, 5.6b). The centromeres are pulled by the spindle fibres towards opposite poles. The attraction between sister chromatids ceases. The reduction in the number of chromosomes is achieved by this separation.

Telophase I

Telophase I does not always occur, but in most animal and some plant cells, the chromatids begin to uncoil and a nuclear envelope forms around each group of chromatids (Figures 5.5c, 5.6c). In many plants, this stage does not occur and the nucleus passes straight into metaphase II of the next stage of division.

There may be a short interphase between the two stages, but there is no replication of the DNA between telophase I and prophase II in either plant or animal cells.

Prophase II

There is no prophase II if interphase is lacking, as the chromosomes will already be condensed. If nucleoli and nuclear envelopes were reformed in telophase I then these will disappear and two new spindles will form at right angles to the plane of the original spindle (Figures 5.5d, 5.6d).

Metaphase II

At metaphase II, each of the chromosomes attaches by its centromere to a spindle fibre and the chromosomes are brought into line along the equators of the spindles (Figures 5.5e, 5.6d).

Anaphase II

Separation of the sister chromatids is achieved at anaphase II (Figures 5.5f, 5.6d). The centromeres divide and the chromatids, now called daughter chromosomes, are pulled by the spindle fibres to opposite poles of the spindle.

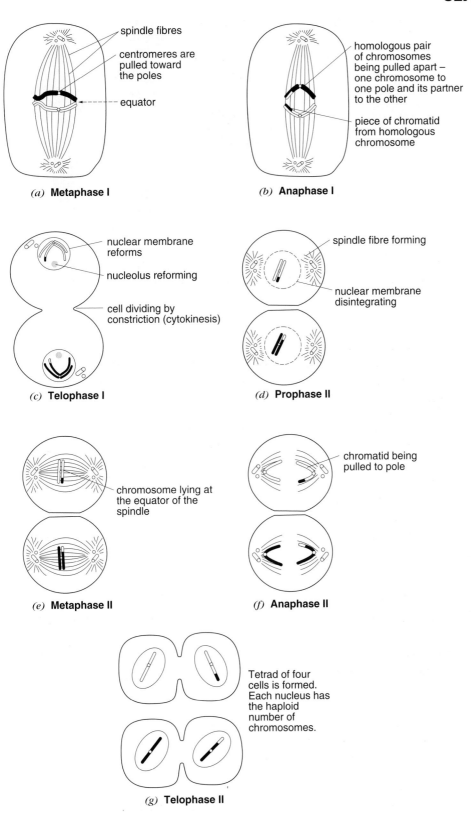

spindle fibres

centromeres are pulled toward the poles

equator

(a) **Metaphase I**

homologous pair of chromosomes being pulled apart – one chromosome to one pole and its partner to the other

piece of chromatid from homologous chromosome

(b) **Anaphase I**

nuclear membrane reforms

nucleolus reforming

cell dividing by constriction (cytokinesis)

(c) **Telophase I**

spindle fibre forming

nuclear membrane disintegrating

(d) **Prophase II**

chromosome lying at the equator of the spindle

(e) **Metaphase II**

chromatid being pulled to pole

(f) **Anaphase II**

Tetrad of four cells is formed. Each nucleus has the haploid number of chromosomes.

(g) **Telophase II**

Figure 5.5 Diagrammatic representations of the remaining phases of meiotic division to form haploid cells: (a) metaphase I, (b) anaphase I and (c) telophase I, followed by (d) prophase II, (e) metaphase II, (f) anaphase II, and (g) telophase II. Plant cells do not have centrioles and a cell plate is formed during cytokinesis

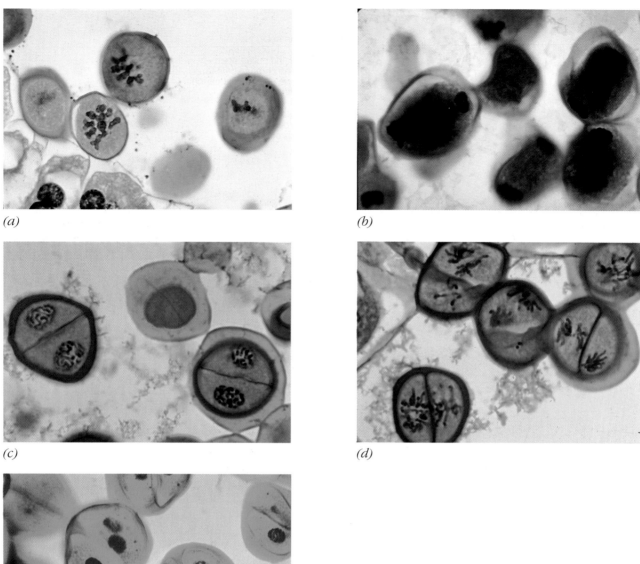

(a)

(b)

(c)

(d)

(e)

Figure 5.6 Photomicrographs of meiosis in Lilium pollen mother cells, showing the remaining phases of division to form haploid cells: (a) metaphase I, (b) late anaphase I in which the cytoplasm becomes very dense and (c) telophase I and first cell cleavage, followed by (d) metaphase II and anaphase II in different cells, and (e) telophase II and second cell cleavage

Telophase II

During telophase II, the daughter chromosomes despiralise and become less visible. The nucleoli reappear and new nuclear envelopes form around the groups of chromosomes (Figures 5.5g, 5.6e).

The first and second telophases are usually accompanied by division of the cytoplasm.

As mentioned earlier, replication of the DNA occurs during interphase before prophase I. At that time, the DNA content of the nucleus doubles. The amount of DNA in each nucleus is half the original amount after the first stage of the meiotic division. By the end of the second meiotic division, the amount in each nucleus is halved again (Figure 5.7).

Significance of meiosis

It is significant that the process of meiosis introduces some genetic variation amongst the offspring of a species. The process of mitosis maintains genetic stability and offers no opportunity for genetic variation except through random mutation. In meiosis, the separation of homologous chromosomes, resulting in the reduction of the diploid (2n) to the haploid (n) number means that each gamete will only carry one form of the gene for a particular characteristic. The crossing over that occurs in prophase I, before the separation of the homologous pairs in anaphase I, results in the exchange of genetic information between maternal and paternal chromosomes, leading to the possibility of new combinations of genes in the gametes. In addition, the orientation of the bivalents on the spindle at metaphase I is completely random, so that the products of the first meiotic division will contain a mixture of chromosomes of maternal and paternal origin. Similarly,

> **DEFINITION**
>
> **Meiosis** is nuclear division in which separation of homologous chromosomes is followed by separation of chromatids, resulting in a reduction of the number of chromosomes in daughter cells.

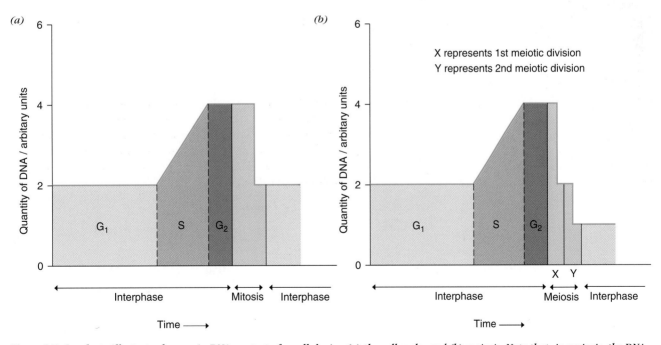

Figure 5.7 *Graphs to illustrate changes in DNA content of a cell during (a) the cell cycle, and (b) meiosis. Note that, in meiosis, the DNA content of each daughter cell is only half that of the parent cell*

during metaphase II, orientation of the pairs of chromatids is random. All these events result in the possibility of a large number of different chromosome combinations in the gametes. This is referred to as **independent assortment of chromosomes** and makes a contribution to the way in which particular characteristics are inherited.

EXTENSION MATERIAL

Mitosis and meiosis compared

There are a number of similarities between the two processes. They both involve:

- replication of the DNA in interphase
- replication of cell organelles
- similar stages of prophase, metaphase, anaphase and telophase, during which similar events take place
- formation of a spindle.

Mitosis and meiosis differ in a number of ways, which are summarised in Table 5.1.

Table 5.1 *Summary of differences between mitosis and meiosis*

Mitosis	Meiosis
consists of one division separating sister chromatids	consists of two divisions: separation of homologous chromosomes followed by separation of chromatids
homologous chromosomes do not associate	homologous chromosomes pair up
no chromomeres visible in prophase	chromomeres often visible in prophase
no crossing over occurs	crossing over occurs
no chiasmata formation	chiasmata formation
daughter nuclei have same number of chromosomes as parent nucleus	daughter nuclei have half the number of chromosomes as the parent nucleus
no genetic variation in daughter nuclei	genetic variation possible in daughter nuclei
can occur in haploid, diploid or polyploid cells (polyploid cells have more than two sets of chromosomes)	occurs in diploid and some polyploid cells
associated with increase in numbers of cells, replacement and repair, asexual reproduction and with gamete formation in plants	associated with gametogenesis in animals and with spore production in plants

Reproduction in flowering plants

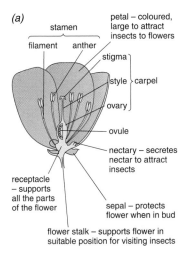

BACKGROUND INFORMATION

Monocotyledons and dicotyledons

Flowering plants are divided into two major subgroups: monocotyledons, having one seed leaf (or cotyledon) in the embryo, and dicotyledons, having two seed leaves (or cotyledons) in the embryo. There are also differences in the internal arrangement of the tissues, but it is usually quite easy to distinguish between the two groups by the appearance of their leaves and flowers.

Monocotyledons usually have:
- long thin leaves with parallel veins
- the flower parts in threes or sixes, i.e. three petals and three sepals.

Often the petals and sepals look the same and they are then called perianth segments. In the grasses (Graminae), a large family of monocotyledons, there are no petals or sepals and the 'flowers' are made up of leaf-like structures called **bracts**. The palea, lodicules and lemma are bracts in grass flowers and are shown in Figure 5.8c.

Dicotyledons usually have:
- a wide range of leaf forms with the veins forming an intricate network
- the flower parts in fours or fives, or multiples of fours and fives
- petals and sepals that are different from each other, i.e. petals often coloured and sepals green.

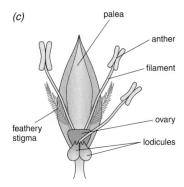

This flower does not have a nectary, but produces masses of pollen which insects collect

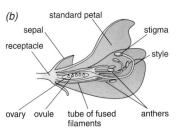

The palea and lodicules are bracts. The lemma, another bract, has been removed. They are green.

Structure and function of the principal parts of flowers

The reproductive organs of flowering plants are contained in structures called **flowers**. Many flowers possess both male and the female structures and are referred to as hermaphrodite. The male organs are the **stamens**. Each stamen consists of a stalk, called the **filament**, bearing the **anthers**. These are lobed and contain four **pollen sacs** in which the pollen is formed. The female organs are called **carpels** and consist of a **stigma**, receptive to the pollen, and a **style** connecting the stigma to the **ovary**. Inside the ovary, one or more **ovules** develop.

The male and female organs develop at the top of the flower stalk, in a region known as the **receptacle**. Surrounding the reproductive organs, there are often **sepals**, **petals** and **nectaries**, arranged in circles, or whorls. Sepals and petals protect the developing flowers and may play a role in the way in which pollination is achieved. Nectaries, which produce a sugary substance called **nectar**, are associated with insect pollination.

Figure 5.8 (a) Generalised structure of a dicotyledonous flower; (b) flower structure in the family Papilionaceae, insect-pollinated; (c) flower structure in the family Gramineae (grasses), wind-pollinated

EXTENSION MATERIAL

Development of pollen and ovules

In the early stages of the development of the pollen sacs of an anther, there is a central mass of cells, called **pollen mother cells**. These cells undergo meiosis to produce groups of four haploid cells, called tetrads. Each cell of a tetrad develops into a **pollen grain**. The haploid nucleus divides by mitosis to produce a generative nucleus and a pollen tube nucleus. The generative nucleus divides once more by mitosis to produce two nuclei that function as the **male gametes**. The pollen grains secrete a thin inner wall, called the intine, and a thick outer wall, the exine. In wind-pollinated species of flowering plants, the exine of the pollen grains is smooth, but in insect-pollinated species it is often sculptured or pitted (Figure 5.9).

(a)

(b)

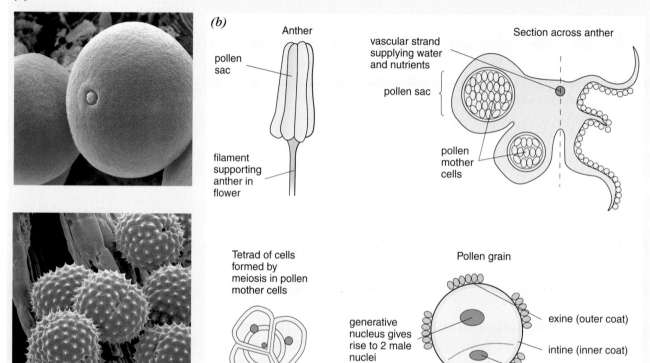

Figure 5.9 (a) Pollen grains from (top) wind-pollinated and (bottom) insect-pollinated flowers; (b) development of pollen grains

As the pollen grains mature, the cells surrounding each pollen sac shrink and break down. Fibrous layers develop in the wall of the anther. As the anther dries out, tensions are set up and the wall splits along a longitudinal line of weakness. The two edges curl away and the mass of mature pollen grains is exposed.

Inside the ovary of the carpel, each ovule begins as an outgrowth of the part of the ovary wall called the placenta. This tiny structure is called the **nucellus**. At the top of the nucellus, a diploid mother cell undergoes meiosis to produce four haploid cells. Usually only one of these haploid cells will continue to develop. It divides three times more by mitosis to produce an **embryo sac** containing eight haploid nuclei. It is in this embryo sac that the female gamete develops. As the nucellus gets bigger, it is surrounded by two layers of cells, called **integuments**, which grow from the base of the ovule. When the ovule is mature, these integuments do not completely surround the nucellus. A small opening, called the **micropyle**, is left.

As the ovule matures, each of the eight nuclei becomes surrounded by cytoplasm and they are arranged in a definite pattern in the embryo sac. The three nuclei at the micropyle end are called the **egg apparatus**. The female gamete, or **egg cell**, is in the centre with cells called **synergids**, one on either side. At the opposite end is another group of three cells, called **antipodal cells**. The remaining two nuclei, called the **polar nuclei**, stay in the centre of the embryo sac. They may stay as separate nuclei, or they may fuse to form a central diploid nucleus. Before pollination and fertilisation, each ovule consists of the nucellus, containing an embryo sac, surrounded by two integuments (Figure 5.10).

Pollination and fertilisation

Before fertilisation can occur, mature pollen grains containing the male gametes must be transferred to the receptive stigma in a process referred to as pollination. In some species, **self-pollination** occurs. Pollen from the anthers is transferred to the stigma of the same flower or another flower on the same plant. In other species, **cross-pollination** is achieved, where pollen from the anthers of one flower is transferred to the stigma of a flower on another plant of the same species.

Pollination is usually achieved by wind or insects transferring the pollen from the anthers to the stigmas. Many flowering plant families, such as the Gramineae (grasses) and many trees, are **wind-pollinated**. This mechanism necessitates the production of vast quantities of light, smooth pollen to maximise the chances of some landing on the mature stigmas of the flowers. **Insect pollination** involves the insect transporting pollen on its body and increases the chances of pollen reaching the stigmas. Insect-pollinated flowers are adapted to this mechanism in that they are coloured and scented to attract the insects, they produce nectar and/or pollen as food for the insects and there are often structural modifications which ensure that pollination is achieved. Table 5.2 summarises the major differences between wind-pollinated and insect-pollinated flowers.

Following successful pollination, the epidermal cells of the stigma secrete a solution of sucrose, which stimulates the germination of the pollen grain. A **pollen tube** grows out through one of the pores in the wall of the pollen grain and rapidly penetrates the tissue of the style. The growth of the pollen tube is controlled by the **tube nucleus**, which is located at the tip of the tube, and involves the secretion of digestive enzymes, allowing the penetration of the tissues. The enzymes soften the outside of the stigma and the middle lamella of the cell walls. Pollen tubes are positively hydrotropic (growing towards water) and negatively aerotropic (growing away from oxygen). As the pollen tube approaches the ovule, it becomes positively chemotropic to a substance produced by the micropyle and so grows in that direction.

The generative nucleus in the pollen grain undergoes mitosis, forming two **male nuclei**. This division may occur before the mature pollen is shed from the anthers or it may occur during the growth of the pollen tube. The tip of the pollen tube enters the ovule through the micropyle and comes into contact with the embryo sac near the site of the synergids. The male nuclei are

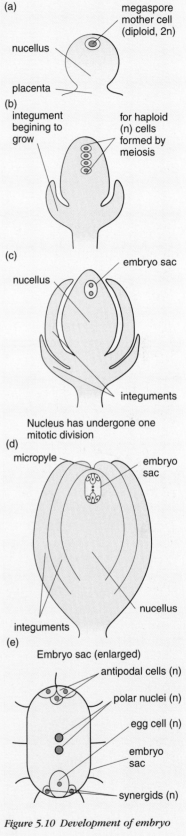

Figure 5.10 Development of embryo sac

Table 5.2 *The differences between wind-pollinated and insect-pollinated flowers*

Feature	Wind-pollinated flowers	Insect-pollinated flowers
position of flowers	above leaves (grasses) or produced before leaves appear (many trees such as hazel, willow)	often, though not always, above leaves; either solitary and large or smaller and in clusters so conspicuous
petals	small, inconspicuous, sometimes absent (grasses); if present, not brightly coloured	large, brightly coloured, conspicuous; attractive to insects
nectaries	absent	present; insects feed on nectar
scent	not scented	often scented; attracts insects
stamens	hanging outside flower (pendulous)	enclosed within flower
anthers	move freely (versatile) so that pollen is easily dispersed	fixed to filaments; positioned so that they come into contact with visiting insect
pollen	produced in large quantities; light, smooth pollen grains	less produced; pollen grains larger, sculptured walls to aid attachment to insects and to stigma
stigma	large, often branched, often feathery, hanging outside flower to trap pollen	small, enclosed within flower; positioned so that it comes into contact with visiting insect

released through a pore which develops in the tip of the pollen tube. One male nucleus will fuse with the egg nucleus to form a **diploid zygote** and the other fuses with the two polar nuclei in the centre of the embryo sac (Figure 5.11). The diploid zygote gives rise to the **embryo** and the nuclei in the centre fuse to form a triploid nucleus, the **primary endosperm nucleus**, which gives rise to the nutritive tissue known as the **endosperm**. This type of fertilisation is called **double fertilisation** as it involves two fusions. It is also worth noting that the transfer of gametes and the process of fertilisation is an adaption to life on land because no water is needed. In the less adapted groups of plants, such as the mosses and the ferns, water is needed for the transfer of gametes from the male to the female organs.

Mechanisms for ensuring cross-pollination

Self-pollination leads to **self-fertilisation**, where the offspring will only show limited variation due to the recombination of genes from a single parent. Self-fertilisation results in **inbreeding**. If an individual plant, heterozygous at a single locus (**Aa**), is repeatedly self-fertilised, the resulting population becomes homozygous at that locus. Two lines, **AA** and **aa** become established. These are called **pure-breeding lines** and mean that there will be no variation occurring at that locus. This can happen for a large number of genes, so natural populations of such plants will show little variation. Self-pollination and

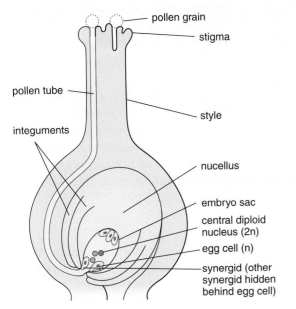

pollen grain

stigma

Tip of pollen tube has opened, releasing male nuclei

pollen tube

One male gamete (n) fuses with the egg cell (n) to give a diploid zygote (2n)

style

integuments

One male gamete (n) fuses with the fusion nucleus (2n) to give a triploid (3n) endosperm nucleus

nucellus

embryo sac

central diploid nucleus (2n)

egg cell (n)

synergid (other synergid hidden behind egg cell)

● male nuclei

Figure 5.11 A mature carpel showing fertilisation

self-fertilisation favours the survival of uncommon or widely-dispersed species. The seeds produced from the fertilised ovules do not depend on wind or insects to disperse the pollen. Common species which show self-pollination are the garden pea (*Pisum sativum*), groundsel (*Senecio vulgaris*) and chickweed (*Cerastium arvense*). In some species, self-pollination occurs if cross-pollination fails.

Cross-pollination, leading to cross-fertilisation and **outbreeding**, occurs in many species and has the advantage of introducing genetic variation, because two different parent plants are involved. Each ovule in a carpel must be fertilised by the male nuclei from separate pollen grains. These pollen grains may come from different parent plants, a feature which has the effect of increasing variation still further. A number of different mechanisms have evolved which prevent self-pollination and improve the chances of cross-pollination. Such mechanisms may involve separation of the male and female reproductive organs, dichogamy, self-sterility (self-incompatibility) or heterostyly.

Although the flowers of most plant species are hermaphrodite, several have separate male and female flowers. **Dioecious** species, such as poplar (*Populus* sp.), willow (*Salix* sp.) and holly (*Ilex* sp.), have male and female flowers on separate plants, so self-pollination is impossible. In **monoecious** species, such as hazel (*Corylus* sp.) and oak (*Quercus* sp.), male and female flowers occur on the same plant, but self-pollination is avoided by a difference in the timing of their development.

In hermaphrodite flowers, self-pollination can be avoided if the male and female organs mature at different times, a condition known as **dichogamy**. In **protandry**, the stamens ripen and the pollen is released from the anthers before the carpels are mature, so the stigmas of the flower are not receptive to pollen at this stage. This mechanism is found in white deadnettle (*Lamium*

QUESTION

List the advantages and disadvantages of inbreeding and outbreeding.

97

SEXUAL REPRODUCTION

*Figure 5.12 Protogyny in ribwort plantain (*Plantago lanceolata*)*

alba), Canterbury bell (*Campanula medium*), rose-bay willow-herb (*Epilobium angustifolium*) and sage (*Salvia* sp.). In the reverse situation, known as **protogyny**, the carpels mature before the stamens, so the stigma of the flower is receptive to pollen before the anthers have ripened and released their pollen grains.

An example of protogyny is seen in the ribwort plantain (*Plantago lanceolata*), where the flowers are arranged on an erect spike (Figure 5.12). The lower flowers open first and initially the stigma is receptive to pollen. Later in the same flower, the anthers ripen and release mature pollen. The mature pollen will always be released at a lower level than the receptive stigmas, so it is unlikely that self-pollination will occur.

Many wind-pollinated flowers, such as members of the Gramineae, are protogynous, but the condition is generally less common than protandry. Among insect-pollinated flowers showing protogyny are wild arum (*Arum maculatum*), bluebell (*Endymion non-scriptus*) and figwort (*Scrophularia nodosa*).

EXTENSION MATERIAL

Self-incompatibility, or **self-sterility**, occurs if pollen from the same flower, or another flower on the same plant, lands on the stigma and fails to germinate. Sometimes germination may occur, but the pollen tube grows very slowly and fails to reach the ovules. Proteins are produced on the surfaces of pollen grains and stigmas. Physiological mechanisms enable the stigmas to distinguish between the proteins on pollen from the same plant and pollen from different plants of the same species. When pollen lands on the stigma of the same plant, the proteins are identical and the pollen fails to germinate, so pollen and stigmas are incompatible. Pollen from a different plant of the same species, carrying different surface proteins, will germinate and is said to be compatible. Different forms of these surface proteins are determined by alleles of an incompatibility gene. This recognition system prevents inbreeding and encourages outbreeding.

QUESTION

Work out why pollen from plant A is needed to pollinate plant B when plant A has alleles S_1 and S_2 at the incompatibility locus and plant B has alleles S_3 and S_4.

ADDITIONAL MATERIAL

Heterostyly describes the situation where different kinds of floral morphology exist within a species. One of the best known examples is seen in the primrose (*Primula vulgaris*), where there are differences in the length of the style and the positioning of the anthers. In the thrum-eyed flower, the style is short, placing the stigma low down in the corolla tube. The anthers are situated high up in the corolla tube, above the stigma. In the pin-eyed type, the anthers are low down the corolla tube, but the style is long and the stigma is above the anthers.

Insects visiting the thrum-eyed flowers have pollen deposited high up on the proboscis as they reach inside the flower for nectar. When a pin-eyed flower is visited, the pollen from the thrum-eyed flower gets brushed against the stigma. Pollen from the pin-eyed flower is deposited about half-way along the proboscis, just in the right position to come into contact with the style of the next thrum-eyed flower visited. However, this mechanism is not particularly effective as, when the insect withdraws from the pin-eyed flower, pollen from the anthers can be drawn up onto the stigma. Similarly, pollen from the anther of the thrum-eyed flower can get taken down to the stigma as the insect enters. Additionally, pollen from the anthers of the thrum-eyed flower can just fall onto the stigma. There is incompatibility between the pollen and the stigmas of the same flower, so self-fertilisation is prevented. Investigation of the pollination mechanism found in the primrose has shown that the development of the pin- and thrum-eyed flowers is controlled by two linked genes determining style length and anther position. These two genes also determine the form of the stigma, pollen grain size and the production of the surface proteins involved with the incompatibility reaction.

QUESTION

Make drawings to show the position of the anthers and stigmas in pin-eyed and thrum-eyed flowers to help you work out what happens at pollination.

Reproduction in humans

Structure and functions of male and female reproductive systems

The reproductive systems in the male and female differ in both their structure and their physiology. The female system produces a gamete, the **ovum** (oocyte), which is fertilised by the male gamete, or **spermatozoon**. The resulting zygote develops and is implanted in the wall of the uterus where it grows and develops until the baby is born after a gestation period of approximately 38 weeks, from fertilisation to delivery.

The **male reproductive system** consists essentially of four major components:
- the **testes**, or male gonads, situated in the scrotum, which are responsible for producing the male gametes, and for secreting male sex hormones.
- a system of **ducts**, including the epididymis and sperm duct, which collect and store spermatozoa from each testis. The ejaculatory ducts join the urethra, through which spermatozoa are expelled.
- glands, including the **seminal vesicles** and **prostate gland**, which secrete a nutritive and lubricating fluid, called seminal fluid, with which spermatozoa are mixed. **Semen** consists of spermatozoa, seminal fluid, mucus and cells which are lost from the lining of the duct system.
- the **penis**, which contains the urethra and erectile tissue. During sexual arousal, this erectile tissue fills with blood and the penis becomes rigid and increases in both length and diameter. The result is termed an erection which enables the penis to function as a penetrating organ during sexual intercourse.

The functions of the **female reproductive system** are to produce gametes, (the ova), to receive the male gametes, to provide a suitable environment for fertilisation and the development of the fetus and to provide a means of expelling the developed fetus during the process of parturition, or birth. The internal organs of the female reproductive system consist of:
- the **ovaries**, which are the sites of both the production of ova and the secretion of the hormones oestrogen and progesterone.
- a pair of **oviducts** (or Fallopian tubes) which convey the ovum from the ovary to the uterus. Fertilisation of the ovum usually takes place in the oviduct.
- the **uterus**, a hollow, pear-shaped, muscular organ. The lining of the uterus, the **endometrium**, undergoes cyclical changes under the influence of the ovarian hormones oestrogen and progesterone. The **cervix** is part of the uterus which projects through the upper part of the vaginal wall.
- the **vagina**, a muscular tube which is adapted both for the reception of the penis during sexual intercourse, and for the passage of the baby out of the mother's body during birth.

The structures of the male and female reproductive systems are illustrated in Figure 5.13.

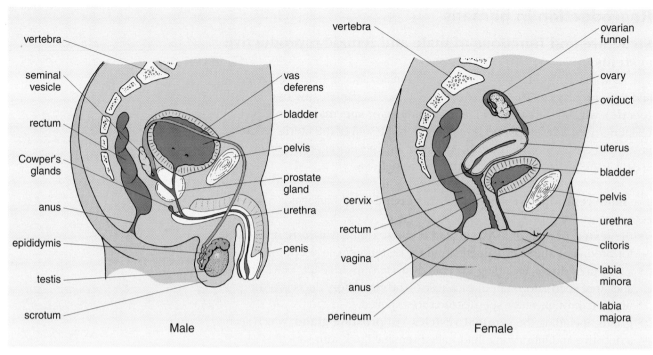

Figure 5.13 The male and female reproductive systems

Production of gametes

Gametogenesis is the process which results in the formation of gametes. It involves a special form of cell division – meiosis – in which the number of chromosomes is halved. The significance of meiosis in sexual reproduction is described on page 91.

Spermatogenesis

Each human testis is packed with numerous coiled **seminiferous tubules**, within which the process of **spermatogenesis** occurs. Spermatogenesis begins at about the time of puberty and normally continues through life. Figure 5.14 shows the major steps in spermatogenesis.

Spermatogonia, also known as germ cells, line the seminiferous tubules and divide by mitosis, giving rise to further spermatogonia (known as spermatogonia Type A) and to spermatogonia which will undergo meiosis to form spermatozoa (Type B). Spermatogonia Type B are also known as **primary spermatocytes** and undergo the first meiotic division to form **secondary spermatocytes**. In humans, this first division takes about 3 weeks to complete. The secondary spermatocytes then rapidly undergo the second meiotic division to form **spermatids**, which undergo a process of development, to form **spermatozoa** (Figure 5.16).

Spermatogenesis occurs in waves throughout the seminiferous tubules; at any give time some areas are active whilst others are at rest. Studies of spermatogenesis in humans have shown that the entire process takes 74 days. Spermatozoa in the seminiferous tubules are not motile, that is, they are

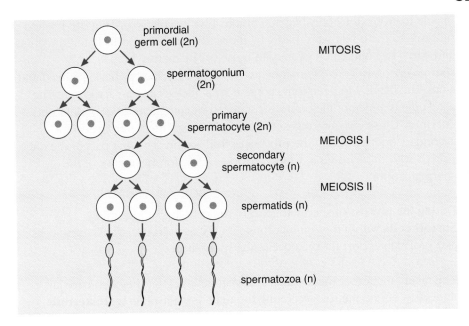

Figure 5.14 Spermatogenesis

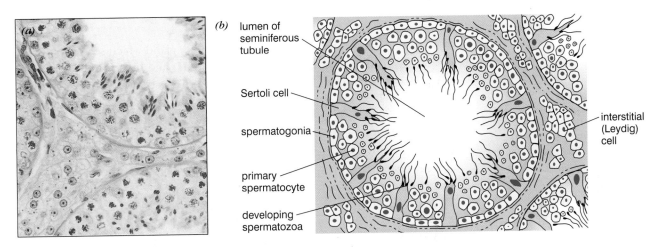

Figure 5.15 Histology of the mammalian testis: (a) transverse section through the testis as seen using a light microscope (high magnification); (b) diagram showing a seminiferous tubule and interstitial (Leydig) cells

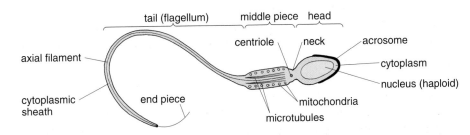

Figure 5.16 Structure of a spermatozoon

unable to swim (they only become motile after ejaculation). They are pushed onwards to the epididymis, partly by the action of cilia on cells lining the ducts and partly by contraction of smooth muscle. The mature human spermatozoon is 55 to 65 μm long and is often described as having a head and a tail. The head, 4 to 5 μm long, consists of a nucleus and an acrosome. The acrosome is a membrane-bound structure containing a number of different enzymes. These enzymes digest a pathway through the outer layers of a secondary oocyte, during the process of fertilisation. The first part of the tail is packed with mitochondria, which provide energy for movement of the spermatozoon.

During the process of spermatogenesis, the cells are supported by **Sertoli cells** within the seminiferous tubules. Sertoli cells act as 'nurse cells' and provide support for the developing spermatozoa.

The main function of the testes is spermatogenesis, but they also have an important endocrine role: secreting the male sex hormone **testosterone**. Testosterone is secreted by cells known as Leydig cells, which are situated in the spaces between the seminiferous tubules. Testosterone controls the rate of spermatogenesis and is responsible for a wide range of male characteristics, including aspects of behaviour, increased growth of muscle tissue and changes in the larynx which cause the voice to 'break' at puberty.

The testes are controlled by the anterior pituitary gland which secretes two hormones: **follicle stimulating hormone (FSH)** and **interstitial cell stimulating hormone (ICSH)**. ICSH is identical in structure to luteinising hormone (LH) which, as described later, has an important role in the menstrual cycle. ICSH stimulates the Leydig cells to secrete testosterone, which has an inhibitory effect on the secretion of FSH and ICSH. In other words, a negative feedback mechanism operates between the anterior pituitary and the testes. FSH acts with testosterone to stimulate the process of spermatogenesis.

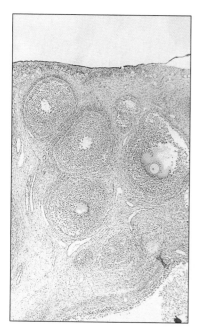

Figure 5.17 Histology of the mammalian ovary. Transverse section as seen using a light microscope (low magnification)

Oogenesis

Oogenesis (Figure 5.18) is the process by which primordial germ cells, known as oogonia, become mature ova. This process begins during early fetal development, when oogonia divide by mitosis. By the fourth and fifth months, some of these oogonia will have enlarged and have the potential to develop into mature gametes. At this stage they are known as **primary oocytes** and begin the first stage (prophase I) of meiosis. By the seventh month of fetal development, the primary oocytes have become surrounded by a layer of flattened follicular cells to form **primordial follicles**. The first stage of meiosis then ceases and no further development occurs until after the female reaches puberty. Then, approximately once a month, a few of the primary oocytes resume meiosis and begin to move towards the surface of the ovary. Usually only one follicle reaches full maturity; the others undergo degeneration. As the follicle develops, it enlarges and fluid begins to accumulate within the follicle. During this stage, the first meiotic division is completed and the oocyte is known as a **secondary oocyte**. At ovulation, the

Figure 5.18 Oogenesis

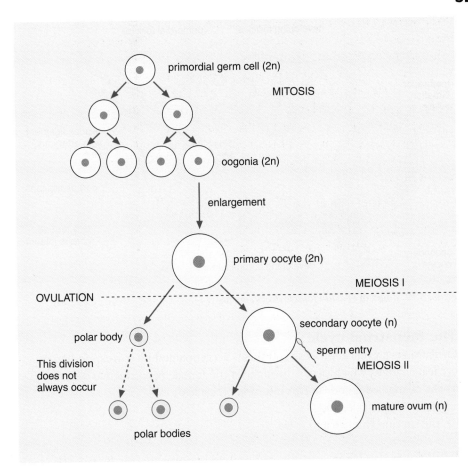

mature follicle (or **Graafian follicle**) ruptures and the secondary oocyte, surrounded by cells from the follicle, is released from the ovary. After the oocyte is released it is usually referred to as an ovum and the second meiotic division will not be completed until the head of a spermatozoon enters during fertilisation.

Notice that during oogenesis the meiotic divisions are unequal, that is, the cytoplasm is not distributed equally between the daughter cells. Only one mature ovum is produced from each primary oocyte, plus three polar bodies, which disintegrate. This process ensures that the ovum has a large store of cytoplasm with all of its organelles and nutrients for early development.

After ovulation, the ruptured follicle fills with a blood clot and cells remaining within the follicle enlarge. This forms a temporary endocrine structure, the **corpus luteum**, which grows for 7 or 8 days. During this time, the corpus luteum secretes **progesterone** and **oestrogen**. If fertilisation and implantation do not occur, the corpus luteum degenerates 12 to 14 days after ovulation to form the functionless corpus albicans (Figure 5.19).

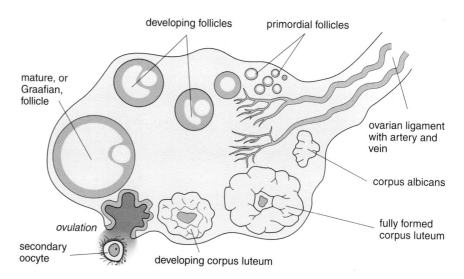

Figure 5.19 Stages of ovarian follicle development

The menstrual cycle

Ovulation is a cyclical process and this is accompanied by corresponding cyclical changes which occur throughout the female reproductive system. These changes depend on two interrelated cycles:

- the ovarian cycle, and
- the uterine (menstrual) cycle.

Both of these cycles are variable, but both last approximately 28 days. The menstrual cycle is controlled by the ovarian cycle, via the hormones oestrogen and progesterone. The ovarian cycle, in turn, is controlled by gonadotrophic hormones secreted by the anterior pituitary gland. The gonadotrophic hormones are **follicle stimulating hormone (FSH)** and **luteinising hormone (LH)** and are secreted in a cyclical pattern.

An initial increase in the blood FSH level stimulates the development of one or more of the primordial follicles and also stimulates follicular cells to secrete oestrogen and small amounts of progesterone. The levels of oestrogen in the blood therefore gradually increase for a few days, then suddenly rise to a peak on about the 12th day of the cycle. About 12 hours after this peak, there is a rise in the levels of both LH and FSH which triggers ovulation. LH also causes the formation of the corpus luteum from the ruptured follicle, which secretes oestrogen and progesterone. If pregnancy does not occur, the lack of FSH and LH causes the corpus luteum to degenerate and the levels of progesterone and oestrogen fall.

The changing concentrations of oestrogen and progesterone during the cycle are responsible for cyclical changes in the uterus. The wall of the uterus has three layers:

- a thin outer layer in contact with the body cavity
- a thick, smooth muscle layer, the **myometrium**
- an inner lining, called the **endometrium**, which provides the environment for development of the fetus.

Figure 5.20 Female reproductive cycles

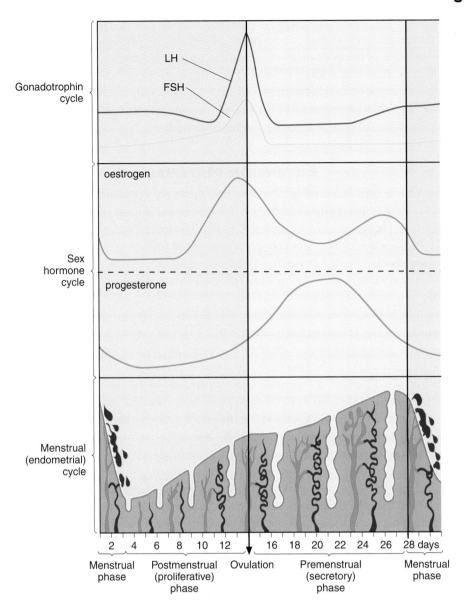

During the uterine cycle, the endometrium undergoes cyclical changes in structure, which can be divided into three phases (Figure 5.20):

- the **menstrual phase**, which occurs on days 1 to about 5 of a new cycle. During this phase, the outer layers of the endometrium are lost (menstruation).
- the **proliferative phase** (or follicular phase), which lasts from about day 6 to day 13 or 14 in a 28 day cycle. During this phase, the endometrium becomes thicker as tissue which was lost in menstruation is repaired.
- the **secretory phase** (or luteal phase) during which glands in the endometrium start to secrete a thick, glycogen-rich mucus. In this phase, which lasts from ovulation to the end of the cycle, the endometrium is prepared for implantation of the fertilised ovum.

As the blood oestrogen levels rise during the proliferative phase, they produce several changes in the endometrium, including repair and thickening, and an increase in the water content of the endometrium. Increasing blood progesterone levels during the secretory phase are responsible for maintenance of the endometrium, secretion by the endometrial glands and a further increase in the water content. The drop in levels of both oestrogen and progesterone, due to degeneration of the corpus luteum towards the end of the ovarian cycle, is responsible for loss of the outer layers of the endometrium which characterises the menstrual phase.

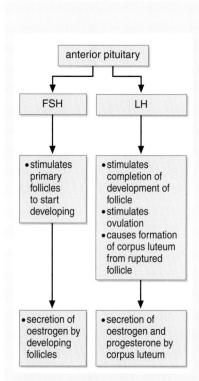

Figure 5.21 Summary of main effects of gonadotrophic follicle stimulating hormone (FSH) and luteinising hormone (LH) on the ovaries

EXTENSION MATERIAL

Feedback and control of the ovarian cycle by gonadotrophins

So far, we have described cyclical changes in the ovary and in the endometrium, and how these changes are, in turn, controlled by the gonadotrophic hormones FSH and LH. The secretion of these gonadotrophins depends on the activity of the **hypothalamus**, which is one reason why female reproductive cycles can be affected by emotional influences, including stress (psychogenic influences).

The hypothalamus controls the secretion of both FSH and LH by means of **gonadotrophin releasing hormone** (GnRH), sometimes referred to as luteinising hormone releasing hormone (LHRH). Both negative and positive feedback mechanisms help to control the secretion of FSH and LH (Figure 5.21). These mechanisms depend on the secretion of oestrogen and progesterone by the ovaries, and the secretion of GnRH by the hypothalamus. Remember that at the start of the menstrual cycle, the blood level of FSH starts to rise, which in turn stimulates the secretion of oestrogen by follicular cells in the ovary. Oestrogen exerts two types of feedback on the secretion of gonadotrophins: at low levels it inhibits secretion (negative feedback), but at relatively high concentrations it stimulates secretion of gonadotrophins (positive feedback). This effect is responsible for the mid-cycle peak in LH and FSH which triggers ovulation. After ovulation, the level of LH remains relatively high, which stimulates the formation of the corpus luteum. Towards the end of the cycle, however, secretion of LH is inhibited as a result of negative feedback exerted by rising levels of progesterone. As the corpus luteum degenerates, the levels of progesterone fall, and the concentrations of FSH gradually increase again. Oestrogen and progesterone probably exert their effects by affecting the secretion of GnRH by the hypothalamus and changing the sensitivity to GnRH of cells in the anterior pituitary.

Fertilisation, implantation and early development

Transfer of spermatozoa intro the vagina involves erection of the penis and ejaculation of semen. Erection occurs as a result of a nervous reflex initiated by various visual, psychological and touch stimuli. As a result of dilation of arteries and arterioles in the penis, spaces in the erectile tissue become distended, which compresses the veins. More blood enters the penis than leaves it through veins and it becomes larger and rigid. Ejaculation of semen is also a reflex involving the same stimuli that initiate erection. Rhythmic contractions of muscles at the base of the penis propel semen into the vagina.

Normal sperms are motile and can move at a rate of about 1 mm per minute. Assisted by muscular movements of the uterus, sperms make their way through

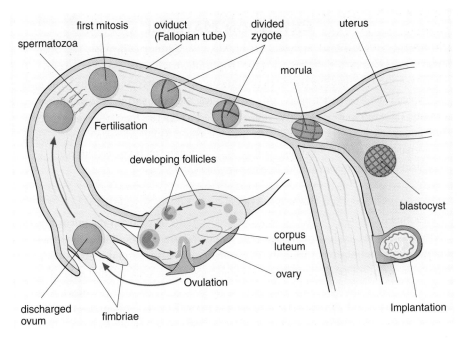

Figure 5.22 Fertilisation and implantation

the cervix and uterus and into the oviducts (Fallopian tubes). Fertilisation occurs most frequently in the outer one third of the oviduct, as shown in Figure 5.22.

Of the 10^8 to 5×10^8 sperms released into the vagina at ejaculation, fewer than 100 reach the oviduct. Only one sperm fertilises the ovum and, as soon as the head of one sperm enters the ovum, complex mechanisms in the ovum are activated to prevent further sperm entry. The 23 chromosomes from the sperm combine with the 23 chromosomes present in the ovum to restore the diploid number of 46 chromosomes.

After ovulation, the ovum lives only for about 24 hours. Sperms may live for up to a few days after entering the female reproductive tract and so sexual intercourse from about 3 days before ovulation occurs, to 1 day after ovulation, may result in fertilisation.

The fertilised ovum is referred to as a **zygote** and it immediately begins to divide, by mitosis, as it travels down the oviduct. In about 3 days, a mass of about 16 cells, known as a **morula**, is formed. The morula enters the uterus and fluid from the uterine cavity enters the morula. The cells rearrange themselves to form a hollow structure, known as a blastocyst. The **blastocyst** begins to implant in the endometrium and, in about 10 days from fertilisation, the blastocyst is completely implanted.

EXTENSION MATERIAL

The blastocyst consists of an outer layer of cells and an **inner cell mass**. The outer wall of the blastocyst is referred to as the **trophoblast** and gives rise to structures which support the embryo during development. As the blastocyst develops further, the inner cell mass forms a structure with two cavities: the **yolk sac** and the **amniotic cavity**. Cells within the yolk sac produce blood cells until this function is taken over by the embryonic liver. The amniotic cavity becomes filled with fluid, mainly derived from maternal blood, which physically cushions the developing embryo, maintains a constant temperature and allows free movement of the fetus.

The inner cell mass develops to form the tissues of the baby itself. Early in development, cells within the inner cell mass form the three **primary germ layers**, referred to as the endoderm, mesoderm, and ectoderm. Cells within each of these germ layers continue to divide and differentiate to give rise to all the tissues, organs and systems of the body. As examples, the endoderm gives rise to the lining of the digestive and respiratory tracts; mesoderm forms most of the skeletal muscles and bones, kidneys and gonads; ectoderm forms the epidermis of skin and various components of the nervous system. During the third week of development, the process of **neurulation** occurs, in which the neural tube is formed, from ectoderm, along the dorsal axis of the embryo. The neural tube is the precursor of the nervous system.

Functions of the placenta

Cells and tissues surrounding the embryo develop to form the **placenta**. The placenta contains many structures known as chorionic villi, which connect the fetal blood vessels to the placenta. Collectively, the chorionic villi provide a large surface area for the interchange of materials between fetal blood and the maternal blood, which are effectively kept separate. Figure 5.23 shows that chorionic villi dip into the maternal blood space, which helps to increase the area of contact between the two circulations. The placenta acts as the site of exchange of many materials between the fetus and the mother, including oxygen and carbon dioxide, nutrients, metabolic waste, antibodies and hormones. The placenta also secretes a number of hormones during pregnancy, such as human chorionic gonadotrophin (HCG), oestrogen and progesterone. During early pregnancy, chorionic gonadotrophin maintains the

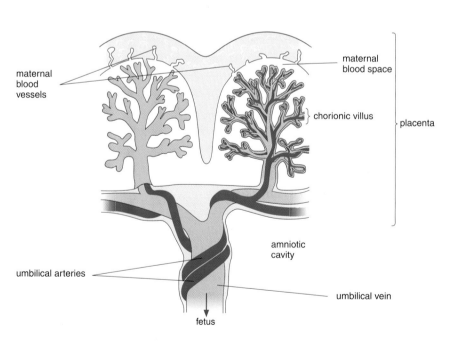

Figure 5.23 Diagrammatic structure of the placenta

corpus luteum in the ovary, which continues to secrete progesterone. Progesterone, in turn, maintains the thickening of the lining of the uterus (the endometrium) and prevents menstruation. As pregnancy progresses, this function of the corpus luteum is taken over by the placenta, which secretes increasing concentrations of oestrogen and progesterone.

The main functions of the placenta may be summarised as:
- keeping fetal and maternal blood separate, while allowing exchange of substances
- exchange of oxygen and carbon dioxide between maternal and fetal blood
- transfer of nutrients from maternal blood to fetal blood
- removal of excretory products, such as urea, from fetal blood
- transfer of antibodies from maternal blood to fetal blood
- secretion of hormones.

The length of pregnancy, about 38 weeks from fertilisation, is divided into three 3-month segments called trimesters. During the first trimester, the term **embryo** is used to describe the developing individual. Towards the end of the first trimester, from about week 8 until birth the term embryo is replaced by the term **fetus**. Stages of embryonic and fetal development are illustrated in Figure 5.24.

Birth and lactation
Birth, or **parturition**, is the point of transition between the prenatal and postnatal periods of life. The term **labour** is used to describe the processes which result in the birth of a baby. Labour is divided into three stages:
- Stage 1: period from the onset of uterine contractions until dilation of the cervix is complete
- Stage 2: from the time of maximal dilation of the cervix until the baby exits through the vagina
- Stage 3: process of expulsion of the placenta through the vagina.

The processes which initiate and control parturition are not fully understood, but can be summarised as follows:
- the sensitivity of the uterine muscle to oxytocin increases towards the end of pregnancy
- in sheep and goats, parturition is triggered by the release of corticotrophin releasing hormone from the fetal hypothalamus, which stimulates the secretion of adrenocorticotrophic hormone (ACTH)
- ACTH acts on the fetal adrenal glands and cortisol is secreted which passes from the fetus to the placenta, where it stimulates oestrogen secretion and inhibits progesterone secretion
- as a result, the synthesis of prostaglandin F2α is increased in the placenta and uterus
- this increases the sensitivity of the uterine muscle to oxytocin
- oxytocin from the mother's posterior pituitary gland stimulates contraction of the uterine muscle until the baby is pushed out through the cervix and vagina.

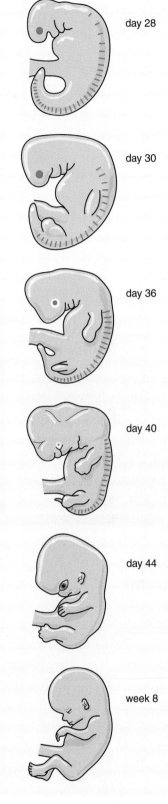

day 28

day 30

day 36

day 40

day 44

week 8

Figure 5.24 Stages of development of the human embryo and fetus

Lactation consists of two processes: milk secretion, and milk ejection. Milk secretion involves the synthesis of milk by secretory cells in the mammary glands. This is stimulated by prolactin, or lactogenic hormone, secreted by the anterior pituitary gland. Milk ejection involves suckling by the baby and a contractile mechanism within the mammary glands which helps to express milk. Suckling stimulates the secretion of oxytocin (and prolactin) by the mother's pituitary gland. Oxytocin is then carried in the blood stream to the mammary glands where it stimulates contraction of cells. This propels milk into ducts where it is accessible for the baby to remove by suckling.

The fluid secreted by the mammary glands during the first 3 days after parturition is termed **colostrum**. This is deep yellow in colour and rich in protein and salts. Milk formed during the first few weeks of lactation is termed transition, or intermediate, milk. Mature milk is produced at the end of the first month. Milk is a rich source of proteins, fat, calcium, vitamins and other nutrients needed by the developing infant. It also provides passive immunity to the baby in the form of maternal antibodies present in the milk.

Passage through the placenta of potentially harmful substances and of viruses

Potentially harmful substances in the mother's blood can cross the placenta and have a number of adverse effects on the developing fetus. Such substances include nicotine, alcohol and heroin. Some viruses are also able to cross the placenta.

- Cigarette smoke contains numerous harmful substances including nicotine, tar and carbon monoxide. Maternal smoking adversely affects the fetus and decreases the chances of survival of the newborn infant, by decreasing the availability of oxygen to the fetus. It has been suggested that oxygen deprivation may be responsible for more than 30 per cent of the deaths of all stillborn infants and is a major cause of intrauterine growth retardation (IUGR).
- Alcohol is freely soluble and easily crosses the placenta. Consumption of alcohol during pregnancy can have tragic effects on the developing fetus. when alcohol enters the fetal blood, the result, called fetal alcohol syndrome (FAS) can cause congenital abnormalities such as microcephaly (abnormally small head), low birth weight, slow physical and mental development of the infant, or fetal death.
- Heroin. If the mother is addicted to heroin, then her baby probably will also be addicted. Such babies are very likely to be born underweight, or prematurely.
- Transmission of infection can also occur across the placenta, an example of vertical transmission. Microorganisms which can be transmitted in this way include the rubella virus, HIV, and the hepatitis B virus. The fetus is particularly susceptible to the rubella virus when maternal infection occurs during the first three months of pregnancy. The virus interferes with the development of the brain, heart, eyes and ears, resulting in a number of malformations, low birth weight, failure to thrive and increased infant mortality.

QUESTIONS

A recent survey in the USA showed that 20 to 25 per cent of women who smoked before pregnancy continued to do so. What are the public health issues associated with smoking (or drinking alcohol) during pregnancy? What factors might influence a woman's decision to continue to smoke during pregnancy?

Development

The postnatal period begins at birth and lasts until death. Although it is often divided into four major periods, it is important to recognise that growth and development are continuous processes which occur throughout life. Gradual changes in the physical appearance of the body as a whole and in the relative proportions of the head, limbs and trunk are particularly noticeable between birth and adolescence. Figure 5.25 shows the changes in the relative proportions of the body parts of a boy from birth to 16 years.

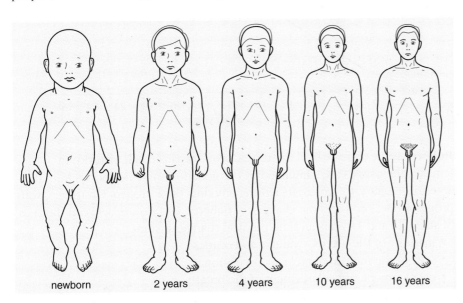

| newborn | 2 years | 4 years | 10 years | 16 years |

Figure 5.25 Changes in the proportions of body parts from birth to 16 years

QUESTION

Describe the changes which occur in the relative proportions of different parts of the body from birth to 16 years.

Briefly, the postnatal periods are as follows:
1 Infancy, which begins at birth and lasts about 18 months.
2 Childhood, which extends from the end of infancy to sexual maturity, or puberty.
3 Adolescence and adulthood. The average age range of adolescence varies, but generally the teenage years (13 to 19) are used. This period is marked by rapid physical growth, resulting in sexual maturity.
4 Older adulthood, characterised by a gradual decline in every major organ system in the body.

A **growth curve** shows the relationship between, say, body mass and the age of the person. Figure 5.26 shows such a growth curve for humans and, after birth, four distinct phases can be recognised:
1 A rapid increase during infancy, especially during the first year.
2 A slower, progressive increase from 3 to about 12 years of age.
3 A marked increase in growth from the time of puberty (the 'adolescent spurt').
4 Even after puberty, there is a slight increase in growth during early adulthood.

Figure 5.26 Growth curve for humans

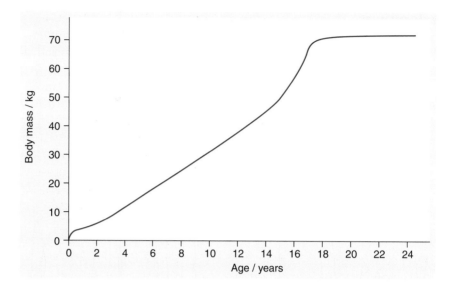

Puberty is the age at which the reproductive organs reach maturity. In females, the ovaries are stimulated by the gonadotrophic hormones from the anterior pituitary, to secrete increasing amounts of oestrogen. Oestrogen, with progesterone, produces the changes of puberty. This occurs between the ages of about 10 and 14 years and the physical changes include:

- maturity of the uterus, oviducts and ovaries
- start of the menstrual cycle and ovulation
- development and enlargement of the breasts
- growth of pubic and axillary hair
- increase in the rate of growth in height
- widening of the pelvis
- increase in the amount of fat deposited in subcutaneous tissue.

Puberty in the male occurs between the ages of about 10 and 14 years. The gonadotrophic hormone ICSH (interstitial cell stimulating hormone) from the anterior pituitary stimulates production of testosterone in the testes. This hormone influences growth and development of the body to maturity. The physical changes which occur during puberty in the male include:

- growth of muscle and bone resulting in a marked increase in height
- the voice 'breaks', or deepens, due to growth of the larynx
- growth of pubic and axillary hair and of hair on the face, chest, abdomen
- enlargement of the penis and scrotum
- production of spermatozoa.

Effects of ageing

Most body systems are in a peak condition and function efficiently during the early years of adulthood. As we grow older, a gradual decline takes place in the functioning of all major organ systems in the body. For example, during older adulthood, the bones undergo changes in their content of calcium (calcification) and texture. With increasing age, changes in calcification may result in a decrease in bone size and in bones which become porous and prone to fracture. Degenerative joint diseases, such as **osteoarthritis**, are common in the elderly. Osteoarthritis is a degenerative change in the cartilage in joints, which in the early stages loses its smooth appearance and becomes flaky. The damaged cartilage is gradually worn away until the underlying bone is exposed. These changes lead to joint pain and reduced movement at the joint. The knees and hips are the most commonly and severely affected.

Degenerative diseases of the heart and blood vessels are a common and serious effect of ageing. Fatty deposits may build up in the walls of blood vessels, resulting in **atherosclerosis**, which may affect all arteries, but the aorta, cerebral and carotid arteries tend to be the most severely affected. The consequences of atherosclerosis include narrowing of the lumen of the artery, which obstructs blood flow, and weakening of the vessel wall.

Ageing also affects the reproductive system. Men may continue to produce gametes as they age, but in women between the ages of about 45 and 60 years, ovulation and the menstrual cycle become less regular and eventually stop. This is known as the **menopause** (Figure 5.27) and results from a decrease in the secretion of sex hormones, particularly oestrogen. Follicle stimulating hormone (FSH) and luteinising hormone (LH) continue to be secreted at higher levels than normal. This results in a number of both physiological and psychological effects, including disturbance of temperature control ('hot flushes') and mood swings including anxiety and loss of confidence. The decrease in oestrogen levels may contribute to a condition known as **osteoporosis**. This is due to a reduction in the density and mass of bone, leading to increased risk of fracture, back pain, weight loss and curvature of the spine. Osteoporosis results in about 60 000 hip fractures a year in the UK, 90 per cent of which occur in people over the age of 50, and 80 per cent of whom are women. Hormone replacement therapy (**HRT**) has been shown to reduce bone loss and to reduce the risk of hip and spinal fractures in postmenopausal women. HRT is administered in the form of tablets or skin patches, containing either oestrogen or a combination of oestrogen and synthetic progesterone.

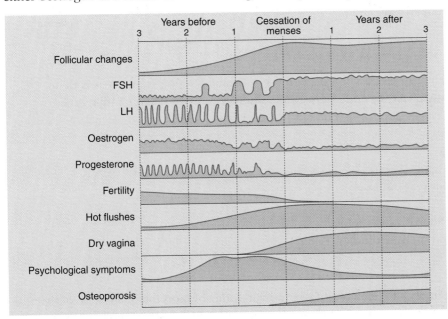

Figure 5.27 Hormonal and other changes associated with the menopause

Practical investigations for Unit 2

Using a simple respirometer

Introduction

A respirometer is used to measure either the volume of carbon dioxide produced or the volume of oxygen consumed by living organisms. There are many different types of respirometer, which vary in complexity from a simple respirometer to elaborate and sensitive types which can be used to measure minute volumes of gas.

One of the problems in respirometry is that gas volumes are influenced by changes in temperature and pressure. Therefore, it is important to control and make allowances for these variables if meaningful results are to be obtained. The respirometer shown in Figure P.1 consists of two tubes; one contains the respiring material or organisms, the other acts as a thermobarometer and compensates for small changes in temperature or pressure. If the air in this tube expands or contracts it will oppose similar changes in the respiration tube. Differences in the level of the manometer fluid are therefore due only to respiratory activity.

If we wish to measure the volume of oxygen used in respiration, the respirometer contains a substance which absorbs carbon dioxide as it is produced. The decrease in gas volume in the respirometer will therefore be equal to the volume of oxygen used. Substances which absorb carbon dioxide include potassium hydroxide and soda lime.

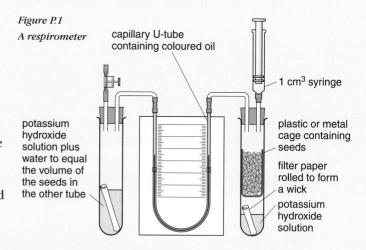

Figure P.1
A respirometer

capillary U-tube containing coloured oil

1 cm³ syringe

potassium hydroxide solution plus water to equal the volume of the seeds in the other tube

plastic or metal cage containing seeds

filter paper rolled to form a wick

potassium hydroxide solution

CORROSIVE
Potassium
hydroxide

CAUTION:

Great care should be exercised when using these substances – potassium hydroxide in particular is corrosive – gloves and safety glasses should be worn when handling it.

Materials

- Respirometer
- Germinating barley grains or pea seeds
- Potassium hydroxide solution, 15 per cent w/v
- Filter paper
- Manometer fluid (Brodie's fluid)
- Water bath and thermometer

Method

1 Carefully pour about 5 cm³ of potassium hydroxide solution into each tube of the respirometer, or add about 5 g of soda lime granules. If you use potassium hydroxide solution, be careful not to allow it to touch the sides of the tubes. If you spill potassium hydroxide solution onto your skin or clothes, wash it off immediately with cold water.

2 Fill the wire or plastic basket with germinating seeds and place in one vessel, ensuring that the seeds do not come into contact with the potassium hydroxide solution.

3 Add to the other tube a volume of water equivalent to the volume of seeds (see Figure P.1).

4 Carefully draw Brodie's fluid into the manometer tube so that it comes about half-way up the scale on either side. It is very important that there are no air bubbles in the manometer fluid.

5 Remove the syringe and screw clip, then connect the manometer to both tubes.

6 Stand the respirometer in a water bath at 20 °C with the manometer outside the water bath.

7 Leave the respirometer for at least 5 minutes to equilibrate. Adjust the piston of the syringe so that it is at about the 0.5 cm^3 mark, then connect to the respirometer as shown in the diagram. Close the screw clip.

8 Use the syringe to adjust the manometer fluid so that the levels are equal on both sides.

9 Record the positions of the syringe piston, the level of the manometer fluid and the time.

10 Record the level of the manometer fluid at suitable time intervals. How frequently you need to take readings will depend on the respiration rate of the organisms you are using.

11 When the manometer fluid reaches the end of the scale, the syringe can be used to return the fluid to its original level.

12 Repeat the experiment at 30 °C and 40 °C, recording your results carefully each time.

13 When you have completed the experiment, record the mass of living material.

Results and discussion

1 Plot a graph of manometer readings against time for each set of results at a particular temperature.

2 If the graph is a straight line, what does it indicate about the rate of respiration at a particular temperature?

3 You can use the syringe to calibrate the respirometer and then calculate the respiration rate. This should be expressed as mm^3 of oxygen used per milligram of living material per hour (mm^3 oxygen mg^{-1} hr^{-1}).

4 What effect did increasing the temperature have on the respiration rate? How could this effect be expressed quantitatively?

PRACTICAL	## Demonstration and measurement of transpiration

Introduction

Measurements of transpiration may be made using a **potometer**. This is an instrument which, strictly speaking, measures the uptake of water by a leafy shoot. However, the volume of water taken up by the shoot is almost exactly the same as the volume lost in transpiration. Relatively small volumes of water are used in photosynthesis, hydrolysis reactions and maintaining the turgor of the cells. Nearly all the water taken up by the shoot is lost in the process of transpiration. In this process water evaporates into the air spaces within the leaves, then water vapour diffuses out of the leaves, mainly through the stomatal pores.

There are many different designs of potometer available, but they generally work on the same principle. Water is taken up by a leafy shoot attached to the potometer and, as water is taken up, an air bubble is pulled along a capillary tube. The rate of movement of the air bubble can be measured against a scale; this corresponds to the rate of transpiration of the shoot.

One type of potometer which is often used in schools and colleges, known as Ganong's potometer, is illustrated in Figure P.2.

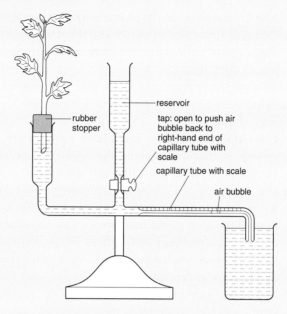

Figure P.2 A potometer set up to measure the uptake of water by a leafy shoot

Materials

- A potometer set up with a leafy shoot
- A stop clock or stopwatch
- An electric fan. If such electrical items are brought in from home, they must be subjected to a portable appliance test to ensure their electrical safety, before they are used in a laboratory
- A small beaker (e.g. 250 cm³)
- Paper towels

Method

There are a number of precautions that need to be taken when setting up and using a potometer, to ensure reliable results.

1 Cut off a leafy shoot from a round, woody stem and immediately place the cut end in a suitable container of water.
2 Back in the laboratory, make a second cut, under water, to remove about 5 cm from the end of the shoot. This helps to avoid air bubbles entering the xylem vessels. When cutting the stem, take care when using scissors or scalpels.
3 Assemble the potometer in a sink full of water. After fitting the shoot through the rubber bung, remove the last 3 cm of bark from the shoot to prevent phloem sap from blocking the xylem.
4 When the potometer has been assembled, the end of the capillary tube should be kept in the beaker of water, as shown in Figure P.2.
5 When you are ready to start recording, introduce an air bubble into the end of the capillary tube, by lifting the tube out of the water. Blot the end carefully with a paper towel and replace the tube in the beaker of water when an air bubble, about 3 to 6 mm long, has entered the capillary tube.
6 When the potometer is set up and the air bubble is moving at a steady rate, record the movement of the bubble along the scale, at suitable time intervals.

7 The bubble can be returned to the beginning of the scale by opening the tap and allowing water to flow out of the reservoir.

8 The experiment can be repeated by exposing the leaves to moving air, using an electric fan.

Results and discussion

1 Tabulate your results showing the times and the position of the bubble along the scale.

2 Plot a graph to show the position of the bubble (vertical axis) against time (horizontal axis).

3 If your graph is nearly a straight line, what does this indicate about the **rate** of water uptake?

4 What effect did moving air have on the rate of uptake of water?

5 If you know the radius of the capillary tube, you can calculate the volume of water taken up, using the formula below.

$$volume = \pi r^2 d$$

where $\pi = 3.142$
r = the radius of the capillary tube
d = the distance moved by the bubble

6 List the factors that affect transpiration and explain *why* each factor affects transpiration.

7 You could find the total surface area of the leaves on the shoot and then calculate the volume of water taken up per unit leaf surface area, per unit time.

PRACTICAL **Stomatal counts**

Introduction

The aim of this investigation is to determine the number of stomata per unit area of a leaf. Some leaves, such as those of *Tradescantia*, can be observed directly with a microscope, but usually the leaf is too thick and it is necessary either to remove a piece of the epidermis, or to make an impression of the epidermis with clear nail varnish. If it is possible to peel the lower epidermis from a leaf, this may be mounted on a microscope slide in a drop of water. Apply a coverslip and examine with a microscope.

Materials

- A microscope
- A bench lamp if the microscope is not fitted with built-in illumination. If using a microscope fitted with a mirror, it is essential never to focus directly on the sun through a window
- Clean microscope slides
- A supply of fresh leaves
- Coverslips
- Clear nail varnish
- Clear adhesive tape
- A stage micrometer

Method

1 Paint a small area (about 5 mm × 5 mm) of the lower epidermis of the leaf *thinly* with clear nail-varnish.
2 When the nail-varnish is completely dry, cover the nail-varnish film with clear adhesive tape and press gently so that the film sticks to the adhesive tape.
3 Peel off the tape and transfer onto a clean microscope slide.
4 Examine your nail varnish impression with a microscope. Focus carefully using the low power objective first, then medium power, before turning to the high power (×40) objective.
5 Count the number of stomata present in the **field of view** (that is, the circular area you can see). Move the slide, and count at least three different areas to find the mean number of stomata present in the field of view.
6 To calculate the number of stomata present per unit area, you need to find the area of the field of view.
7 Use a **stage micrometer** to measure the diameter of the field of view for the objective you used when counting. Focus on the stage micrometer very carefully, as they are expensive.
8 Record the diameter of the field of view.

Results and discussion

1 Record your results in a table, showing the separate counts and the mean number of stomata per field of view.
2 Calculate the area of the field of view using the formula

$$\text{area} = \pi r^2$$

where $\pi = 3.142$
r = the radius of the field of view

As an example, suppose that the **diameter** of the field of view is 420 μm. 420 μm is equal to 0.420 mm, so the **radius** is 0.21 mm. The area of the field of view is therefore

$$\pi \times (0.21)^2$$

$$= 0.14 \text{ mm}^2$$

If you counted 16 stomata in this area, the number per mm^2 will be

$$16 \div 0.14 = 114 \text{ stomata per mm}^2.$$

3 You could investigate the numbers of stomata on leaves of different species of plants, or relate the number of stomata to the rate of transpiration.

PRACTICAL | **Examination of stained blood films and the identification of cells**

Introduction

The aim of this practical is to examine prepared, stained blood films and to identify the different types of cells present. Slides have been prepared already, on which a small drop of blood is spread evenly over the slide, to produce a thin film. This is then allowed to dry, and the cells are stained using, for example, Giemsa, or Leishman's stain. These stain the nuclei, and any granules within the cytoplasm, of white blood cells.

Blood cells are relatively small (human red cells have a mean diameter of 7.3 μm) and it is therefore helpful to view the cells using an **oil immersion objective**. This type of objective lens gives a high magnification (typically $\times$100), but if this is not available, cells can be seen clearly using the $\times$40 objective lens.

Materials

- Microscope fitted with oil immersion objective lens, if possible
- Immersion oil
- Bench lamp if the microscope is not fitted with built-in illumination. If using a microscope fitted with a mirror, it is essential never to focus directly on the sun through a window
- Prepared stained blood film
- Lens tissues

Method

1 If using an oil immersion objective, place the slide on the microscope stage, ensuring that the blood film is on the top surface of the slide. Rotate the eyepieces until the oil immersion objective clicks into place.
2 Place **one drop** of immersion oil directly onto the slide and, looking from the side of the microscope, carefully rack down the objective until the lens just dips into the oil.
3 Look down the eyepiece and carefully continue to focus using the fine focus control until the cells are in focus.
4 After use, carefully wipe the oil from the lens and microscope slide using a clean lens tissue.
5 If you are using a microscope without an oil immersion lens, focus using the low power objective first, then medium power, before changing to the high power ($\times$40) objective.

Results

1 Refer to photomicrographs and diagrams of blood cells to help you to identify the different types of cells present.
2 As a guide, you may find the information in the following table helpful.

Table P.1 *Guide to the identification of blood cells seen in a stained blood film*

Type of blood cell	Description
erythrocyte (red blood cell)	Usually described as 'biconcave discs', these appear as circular structures, with a lighter centre where the cells are thinner. Mammalian erythrocytes have no nuclei. Erythrocytes outnumber white blood cells by about 600 to 1.
neutrophils	Approximately 70% of all white blood cells are neutrophils. They have an irregular, lobed nucleus, which usually stains blue or purple. These cells have fine granules in the cytoplasm.
lymphocytes	These are the next most abundant type of white blood cells and have a relatively large, round nucleus with clear cytoplasm. In small lymphocytes, most of the cell is occupied by the nucleus.
monocytes	Monocytes are the largest of the white blood cells and have a more or less horseshoe shaped, or indented nucleus. The cytoplasm usually appears pale blue or clear.
eosinophils	These cells have large, red-stained granules in their cytoplasm and a lobed nucleus, usually with two lobes.
basophils	These are the least numerous type of white blood cells (typically less than 1%). They have darkly stained granules in their cytoplasm and a lobed nucleus, although the granules often hide this.

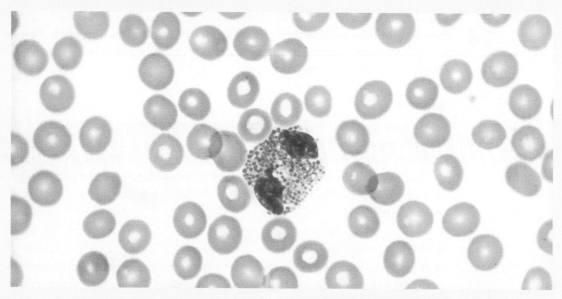

Figure P.3 Photomicrograph of blood cells in a stained smear of human blood. The majority of cells are erythrocytes (red blood cells); the cell in the centre of the picture is an eosinophil (oil immersion objective).

PRACTICAL

Factors affecting the growth of pollen grains

Introduction

Pollen grains can be germinated *in vitro* by placing them in a suitable solution. Pollen grains of some species of plants may germinate in water, but the percentage germination is low, the pollen tubes grow slowly and may burst. The percentage germination and growth of the pollen tube can be increased by using sucrose solution. Sucrose acts as a nutrient and reduces the osmotic effects, so preventing the tubes from bursting. Boron appears to be essential for successful pollen tube growth in some species, so boric acid (trioxoboric acid) is added to the culture medium. The aim of this experiment is to investigate pollen tube growth by culturing pollen grains in sucrose solutions.

Materials

- Flowers, such as *Lilium, Narcissus* or tulips
- Microscope slides and coverslips
- Sucrose solutions: 0.2, 0.3, 0.4, and 0.5 mol per dm^3 made up in distilled water containing 0.01 g per dm^3 boric acid (trioxoboric(III) acid)
- Plasticine
- Small paint brush to transfer pollen
- Disposable 1 cm^3 polythene pipette, or similar
- Microscope
- A bench lamp if the microscope is not fitted with built-in illumination. If using a microscope fitted with a mirror, it is essential never to focus directly on the sun through a window

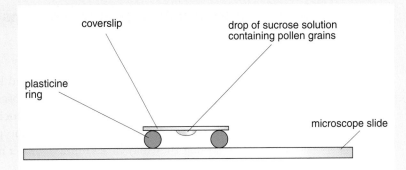

Figure P.4 A hanging drop preparation to investigate the growth of pollen grains

Method

1 Place a ring of plasticine, about 2 cm in diameter, on a microscope slide so that the coverslip will be supported about 4 mm above the surface of the slide (see Figure P.4).
2 Place one drop of sucrose solution in the centre of a coverslip, and dust a small amount of pollen onto the sucrose solution.
3 Carefully invert the coverslip onto the ring of plasticine, so forming a **hanging drop preparation** (Figure P.4).
4 Leave the preparation at a temperature of about 20 to 25 °C for 1 to 2 hours, then examine using a microscope.
5 Count the total number of pollen grains visible in the field of view and the number which have germinated, that is, have a pollen tube.

Results and discussion

1 Record your results in a table, showing the percentage of pollen grains that have germinated in each sucrose solution.
2 Plot a graph to show the relationship between the percentage germination of pollen grains and the concentration of sucrose. Describe the relationship.

Suggestions for further work – opportunities for individual investigations

1 Investigate the germination of pollen grains in sucrose solutions with and without boric acid.
2 Investigate the effect of different concentrations of boric acid on germination of pollen grains. It is suggested that concentrations of boric acid between 0.001 and 0.01 per cent, in distilled water, should be used.
3 Use an eyepiece graticule to measure the **rate** of growth of pollen tubes, measuring the length of the tubes every 30 minutes for three hours.
4 Compare the percentage germination of pollen grains in sucrose solutions for a range of different plant species.

Observations on preparations of an insect testis squash

Introduction

The testes from insects such as grasshoppers and locusts make suitable material for the observation of stages of meiosis. Prepared microscope slides of insect testis squash are available, which have been stained with a dye, such as orcein, to make the chromosomes visible. The testes are gently squashed during the preparation; this spreads the cells out into a thin layer, so that they may be examined with a microscope.

Materials

- Microscope
- A bench lamp if the microscope is not fitted with built-in illumination. If using a microscope fitted with a mirror, it is essential never to focus directly on the sun through a window
- Prepared microscope slide of insect testis squash, such as locust (*Locusta*) or grasshopper (*Chorthippus*)

NB: If these slides are not available, stages of meiosis may also be observed in, for example, preparations of *Lilium* anthers.

Method

1 Set up the microscope and place the slide on the stage.
2 Carefully focus on the preparation using the low power (×4) objective lens.
3 When the cells are in focus, change to the next magnification objective (e.g. ×10) and focus carefully. Look for cells in which the chromosomes are clearly visible.
4 Finally, change to the high power objective lens (×40) and focus using the fine focus control.

Results

You should be able to see the chromosomes clearly and, if the cells are in prophase I or metaphase I, it should be possible to see bivalents (pairs of homologous chromosomes) and chiasmata.

1 Make annotated drawings to show representative stages of meiosis.

Measuring vital capacity

Introduction

The vital capacity (maximum volume of air which can be expired following maximal inspiration) of a person can be measured simply by using a suitably calibrated glass bell jar, supported in a sink of water (Figure P.5).

Materials

- Large (5 dm^3) calibrated bell jar
- Wide diameter rubber or PVC tubing
- Suitable supports for the bell jar

Estimated vital capacities
For males: 2.6 dm^3 m^{-2} body surface area
For females: 2.1 dm^3 m^{-2} body surface area

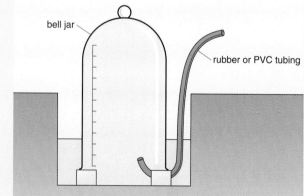

Figure P.5 Method for determining the vital capacity

Method

1 First calibrate the bell jar by inverting it and pouring in known volumes of water. Use a marker pen to graduate the jar.
2 Fill the jar and invert into a large sink filled with water. Support the jar on suitable blocks.
3 If available, use a standing waste so that the sink is about two-thirds full.
4 Each student then uses the tubing after disinfecting with a suitable antiseptic solution and, following maximal inspiration, exhales as far as possible into the jar.
5 The vital capacity can then be recorded.

Results and discussion

NB: When investigating physiological parameters, it should be noted that there is always variation from person to person. The data should therefore be interpreted with care as there are many factors which can influence the results obtained.

1 Record class results in a suitable table.
2 Is there a consistent difference between the vital capacity of males and females? If so, can this be quantified?
3 Investigate the effect of posture on vital capacity and suggest reasons for the results.
4 If available, use a surface area nomogram to determine your body surface area and, using the relationship in the margin above, calculate your estimated vital capacity. Suggest reasons for any differences there may be.

NB: Vital capacity can also be measured conveniently using a pocket spirometer or lung volume bags. These are available from Philip Harris Education.

PRACTICAL	**Variation in breathing rates with physical activity**

Introduction

The aim of this practical is to investigate the effect of physical activity on breathing rate. Although quite simple in principle, this practical can yield quantitative data showing, for example, the effect of incremental exercise on breathing rate.

NB: In any investigations on human subjects, it is essential that the subjects are sufficiently fit and willing to participate. The activities must not be allowed to degenerate into a competition. Asthmatics may need to use their inhalers before taking exercise. Teachers should guard against students taking part, despite knowing there is an underlying medical problem, because they do not wish to be seen as different by their peers.

Materials

- A stopwatch (or stop clock)
- A stepping bench, or equivalent, 41 cm high (although this is not essential for this investigation). This must be strongly constructed, so that it does not collapse. Where such a bench is not available, it is essential that students do not step on and off bench stools, as these are unstable. Other types of exercise may be performed safely.

Method

1 Record your resting breathing rate, by counting the number of times you breathe in and out in 1 minute. It is sometimes difficult to obtain a reliable figure, because breathing rate (and depth) can be consciously altered. Take several readings and obtain a mean value. Breathing rate is expressed as the number of breaths per minute.

2 Now take some form of standardised exercise, such as stepping on and off a bench at a regular rate for 1 minute. Immediately after the exercise, record your breathing rate.

NB: You are not expected to overexert yourself – gentle exercise is all that is necessary in this investigation!

Results and discussion

1 Record your results in a table, showing your breathing rate at rest, and after the period of exercise.

2 Explain your results as fully as you can.

Further work

You could investigate the effect of increasing levels of activity on breathing rate. For example, you could take standard exercise as above for progressively longer periods, such as 1 minute, 2 minutes, and then 3 minutes. You should allow time for your breathing rate to return to its normal, resting value between each period of exercise.

Plot a graph to show the relationship between breathing rate and the level of exercise.

PRACTICAL

Quantitative comparisons between inspired and expired air

SAFETY:
Wear eye
protection
Goggles or face
shield, not just
safety spectacles

CORROSIVE
Potassium
hydroxide

Alkaline
pyrogallol

Introduction

This practical activity enables you to analyse samples of inspired and inspired air in order to calculate the percentages of oxygen and carbon dioxide. However, the method involves the use of very hazardous reagents, potassium hydroxide and alkaline pyrogallol. *It is essential, therefore, that the practical is carried out by a teacher as a demonstration.*

Materials

- Eye protection
- Access to a sink, or large plastic bowl, filled with cold water
- A gas burette, fitted with a Suba-seal stopper (available from Philip Harris Education)
- A 1 cm^3 plastic syringe, fitted with a needle. HANDLE THIS WITH GREAT CARE and ensure that students cannot gain access to this and the solutions
- Potassium hydroxide solution (saturated)
- Freshly prepared alkaline pyrogallol solution (benzene-1, 2, 3-triol, plus potassium hydroxide). To prevent oxygen from the air being absorbed, it is customary to have a layer of liquid paraffin over the alkaline pyrogallol solution

Method

1 Obtain a sample of **expired air** by removing the Suba-seal stopper from a **clean** gas burette and breathe out steadily through the tap (or remove the clip and breathe through the tubing) into the burette. Whilst continuing to breathe out, place the open

end of the burette under the surface of the water in the sink (or bowl). When you have breathed out as far as possible, allow a few cm^3 of water to flow into the burette.

2 Close the tap and replace the stopper, whilst keeping the end of the burette under water.

3 Leave the burette completely submerged in the water for 5 minutes, to allow the air sample to reach the same temperature as the water. Hold the burette so that the levels of water inside and outside are about the same, and then open the tap. Carefully raise or lower the burette until the two water levels are the same again, then close the tap.

4 Raise the burette to eye level, then read and record the volume of gas in the burette. This is the **initial volume**.

5 Carefully inject about 0.5 cm^3 of potassium hydroxide solution through the stopper, whilst holding the burette under water.

6 Keep the burette under water and rock it carefully to mix the air and the potassium hydroxide solution.

7 After at least 5 minutes, level the solution in the burette with the water in the bowl, as in steps 3 and 4.

8 Read and record the volume of gas in the burette.

9 Repeat steps 5 to 8, but this time injecting 0.5 cm^3 of **alkaline pyrogallol**.

10 Again, read and record the volume of gas in the burette.

11 To obtain a sample of **atmospheric air**, fill a gas burette with cold water. Open the tap and allow about 10 cm^3 of water to run out. Close the tap and immediately replace the Suba-seal stopper.

12 To analyse this air sample, follow the method from step 3.

13 Record the results in tables like these shown below.

Table P.2 *Finding the percentage of carbon dioxide*

Gas burette reading	Volume / cm^3
initial gas volume (after step 4)	
volume after adding potassium hydroxide (after step 8)	
difference in volume	

Percentage of carbon dioxide = (difference in volume ÷ initial volume) × 100

Table P.3 *Finding the percentage of oxygen*

Gas burette reading	Volume / cm^3
volume after adding potassium hydroxide (after step 8)	
volume after adding alkaline pyrogallol (after step 10)	
difference in volume	

Percentage of oxygen = (difference in volume ÷ initial volume) × 100

NB: When analysing atmospheric air, there may be an insignificant change in volume after adding potassium hydroxide solution.

Effects of physical activity on pulse rate

Introduction

The aim of this practical is to investigate the effect of exercise on pulse rate. As with all investigations involving human subjects, it is essential that they are sufficiently fit and willing to participate. The activity must not be allowed to degenerate into a competition. Gentle exercise is all that is needed; you are not expected to over exert yourself! Asthmatics may need to use their inhalers before taking exercise. Teachers should guard against students taking part, despite knowing there is an underlying medical problem, because they do not wish to be seen as different by their peers.

Materials

- A digital pulse rate monitor, or heart rate monitor (although not essential for this investigation)
- A stopwatch or stop clock
- An exercise cycle, or stepping bench 41 cm high (if available, but not essential)

Method

1 Record your resting pulse rate. If digital pulse monitors are not available, there is nothing wrong with the 'fingers on the pulse' method, using your index finger and third finger gently pressed on the radial artery in your wrist (see Figure P.6).

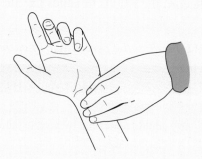

Figure P.6 Determining your pulse rate

2 Perform some standardised exercise, such as stepping on and off a bench at a fixed rate for 1 minute.
3 Immediately after the exercise, record your pulse rate, then again after a rest of 5 minutes.

Results and discussion

1 Record your results in a table.
2 Explain your results as fully as you can.

Suggestions for further work

Record your pulse rate before, during, and after a period of exercise. Plot a graph of your results to show how your pulse rate changed during the investigation

Investigate the effect of incremental activity, such as cycling at increasing speeds for a fixed time, on your pulse rate. You must allow sufficient recovery time between each bout of exercise. Describe the relationship between the level of activity and pulse rate.

Microscopic examination of the histology of ovary and testis

Introduction

The term **histology** means the study of tissues. The aim of this practical is to familiarise you with the structure of the ovary and testis. This will help you to understand the functions of these organs, in terms of both the production of **gametes** by the process of meiosis, and their function in the production of **hormones**. You may find that reference to photomicrographs and to diagrams will help you to interpret the structures as seen with a microscope.

Materials

- A microscope
- A bench lamp if the microscope is not fitted with built-in illumination. If using a microscope fitted with a mirror, it is essential never to focus directly on the sun through a window
- A prepared slide of mammalian ovary (transverse section)
- A prepared slide of mammalian testis (transverse section)

Method

1 Set up the microscope and place the slide of **ovary** on the stage. Focus with the low power (×4) objective lens. You should be able to see **follicles**, in various stages of development, including immature follicles and secondary follicles. It is possible that the section may show one (or more) mature, or **Graafian follicles**, containing the relatively large secondary oocyte. There may also be a **corpus luteum**. This is formed after ovulation and contains cells that secrete the hormones **oestrogen** and **progesterone**. The follicles and other structures within the ovary are surrounded by connective tissue (the stroma) and many blood vessels.

2 Make a labelled, low-power plan drawing of the ovary, and a flow chart to show the stages in the development of a human ovum. On your flow chart, show where mitosis, meiosis I and meiosis II occur.

3 Now examine the slide of testis. At low magnification, you will see many **seminiferous tubules** in section. **Spermatozoa**, the male gametes, are produced within the seminiferous tubules, which are lined with cells called **spermatogonia**. The seminiferous tubules also contain specialised cells known as **Sertoli cells**, which support the developing spermatozoa. Sertoli cells have a distinctive nucleus with a conspicuous, darkly stained nucleolus. The nucleus often appears to be triangular, or oval in shape. In the spaces between the seminiferous tubules, there are small clusters of cells known as **interstitial cells** (or Leydig cells), which secrete the hormone testosterone.

4 Make a labelled, low-power plan drawing to show the structure of the testis, and a flow chart to show the stages in the formation of spermatozoa. On your flow chart, show where mitosis, meiosis I and meiosis II occur.

5 Write a brief account to explain the importance of meiosis in the production of gametes.

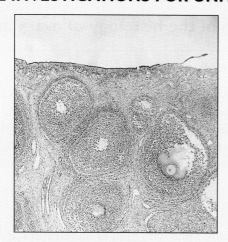 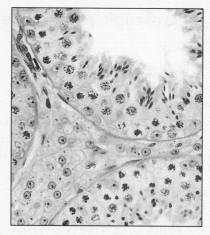

Figure P.7 Photomicrographs to show the structure of ovary (low magnification) and testis (high magnification).

Energy and the Environment

Bluebells in a woodland

Modes of nutrition

6

Autotrophic and heterotrophic nutrition

All living organisms need energy for growth and maintenance. **Autotrophic** organisms are able to use external sources of energy in the synthesis of their organic food materials. Algae, green plants and certain prokaryotes can obtain their energy directly from the sun's radiation and use it to build up essential organic compounds from inorganic molecules. These organisms are **photosynthetic** and possess special pigments that can absorb the necessary light energy. A few specialised prokaryotes are able to use energy derived from certain types of chemical reactions in the synthesis of organic molecules from inorganic ones. These organisms are **chemosynthetic** and include the nitrifying bacteria *Nitrosomonas* and *Nitrobacter*, which are important in the nitrogen cycle (see Chapter 7). In an ecosystem, the photosynthetic organisms are known as **primary producers** in the food chains as they synthesise the organic molecules required by the other organisms, which are known as **consumers**.

(a)

(b)

(c)

Figure 6.1 All these organisms are heterotrophic: (a) lionesses feeding at a kill (carnivore, secondary consumer); (b) bracket fungi (parasites); (c) hermit crab with anemone attached to its shell (mutualism)

All other organisms are **heterotrophic** and need to be supplied with food consisting of complex organic molecules, which they use to obtain energy for metabolism and the materials required for growth, repair and replacement of tissues. These organic molecules are obtained either directly from green plants or from organisms that have fed on green plants. The ways in which heterotrophic organisms obtain their food vary greatly (Figure 6.1), but once obtained, the complex organic compounds have to be broken down into simpler, soluble molecules before they can be absorbed.

Heterotrophic organisms are the consumers in food chains within an ecosystem. Primary consumers are **herbivores**, feeding directly on green plants. Secondary consumers are **carnivores** and feed on the herbivores. Tertiary consumers are also carnivores as they feed on the secondary consumers. Heterotrophic organisms that feed on both plant and animal material are known as **omnivores** (Figure 6.2).

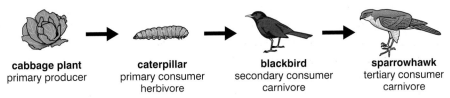

cabbage plant	caterpillar	blackbird	sparrowhawk
primary producer	primary consumer	secondary consumer	tertiary consumer
	herbivore	carnivore	carnivore

Figure 6.2 Food chain showing the relationships between producers and consumers

Four main types of heterotrophic nutrition are recognised:
- **holozoic**, in which complex food is taken into a specialised digestive system, broken down into smaller pieces, and absorbed – this type of nutrition is characteristic of free-living animals
- **saprobiontic/saprotrophic**, in which organisms feed on dead organic remains of other organisms
- **parasitic**, in which an organism obtains food from another living organism, called the **host**
- **mutualism**, a form of **symbiosis**, in which there is a close association between two organisms, each contributing to and benefiting from the relationship.

Holozoic nutrition

Holozoic nutrition is characteristic of higher animals, including humans, and consists of five stages:
- **ingestion** – food is taken into the body through the mouth
- **digestion** – the food is first mechanically broken down by the teeth in the mouth and then chemically broken down by **hydrolysing enzymes** in the **stomach, duodenum** and **ileum**
- **absorption** – the smaller soluble molecules are taken up into the **bloodstream** from the **duodenum** and **ileum**
- **assimilation** – the absorbed products of digestion are incorporated and used by the body
- **egestion** – the undigested parts of the food are eliminated from the body through the **anus** during **defaecation**.

Herbivores

Cattle and sheep are herbivores (Figure 6.3). Their **ruminant** digestive system allows a high proportion of the cellulose in fibrous foods to be digested and become available as energy. They also benefit from features of their nitrogen

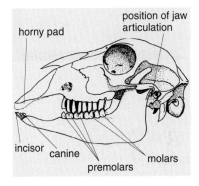

Figure 6.3 Side view of skull showing dentition of a ruminant such as a sheep, illustrating typical herbivore features.
- *small sharp incisors and canines on lower jaw – cut against pad of gum for cropping*
- *diastema (gap between incisors and molars) – for manipulation of food bolus*
- *premolars and molars – for grinding*
- *ridges of enamel – provide efficient cutting edge as grinding surfaces of premolars and molars get worn down*
- *open roots – allow for continuing growing of the tooth*
- *loose jaw articulation – allows sideways movement of lower jaw against upper jaw (related to cud chewing)*

Dental formula $i \frac{0}{3}$ $c \frac{0}{1}$ $pm \frac{3}{3}$ $m \frac{3}{3}$

(i = incisors, c = canines, pm = premolars, m = molars. The line separates the upper and lower jaw)

metabolism and their ability to synthesise water-soluble vitamins. The ruminant alimentary canal is more complex than that of humans due to three additional compartments, the **rumen**, **reticulum** and **omasum**, which come before the true stomach, known as the **abomasum** (Figure 6.4).

In a cow, after the food has been swallowed it passes to the rumen for up to 30 hours. Here it is mixed mechanically and fermented by populations of microorganisms. Coarse material is regurgitated into the mouth, rechewed, then swallowed again back into the rumen. Chewing the cud in this way may continue for up to 8 hours in a day if the diet is very fibrous. The microorganisms in the rumen are mainly bacteria but also include protoctists and yeasts. These microbes become established soon after birth when the calf begins to pick up solid food. The microorganisms digest carbohydrates, particularly polysaccharides with β-links, and this contributes to the breakdown of cellulose. The resulting hexoses are further broken down, by fermentation under anaerobic conditions, to short-chain organic acids (ethanoic, propionic and butyric) with the release of the gases carbon dioxide and methane.

> **example of rumen fermentation reaction**
>
> $$C_6H_{12}O_6 \rightarrow 2CH_3COOH + CO_2 + CH_4$$
>
> glucose   organic   carbon   methane
> acid   dioxide

Energy released in these reactions is utilised by the microorganisms for their own metabolic activities. The gases escape from the animal by belching and are wasted. The acids, known as volatile fatty acids (VFAs) are absorbed through the walls of the rumen and contribute to the energy requirements of the animal.

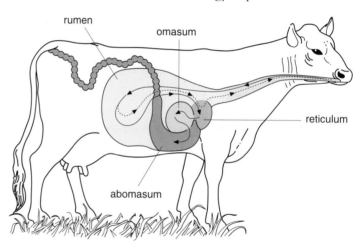

Figure 6.4 Ruminant digestive system in a cow– the arrows show the pathway taken by food, including chewing the cud
- *the **rumen** is the chamber in which fermentation of food material takes place. Capacity is about 150 litres. Microorganisms in the rumen synthesise vitamins of the B-complex*
- *the **reticulum** has no digestive function – concerned with the passage of boluses of food from the rumen to the oesophagus and digested material from the rumen to the omasum*
- *the **omasum** – main function is to remove water and organic acids from digested material passed from the rumen; produces no digestive secretions. Capacity is about 15 litres.*
- *the **abomasum** functions as a 'true' stomach, secreting gastric juices. Digestion here and in the rest of the alimentary canal is similar to that in humans*
- *fibre slows down the passage of food in the gut and adequate amounts in the diet are essential to maintain rumen function.*

Carnivores

Carnivores are meat eaters and their teeth are adapted for catching and killing their food, as well as for shearing the flesh, cutting through tendons and ligaments, and crushing bones. The teeth of the carnivores show many adaptations to their diet (Figure 6.5), including:

- sharp incisors, which can hold the prey and also remove flesh from bones
- large, often curved, pointed canines which assist in holding prey and tearing flesh
- the fourth premolars on each side of the upper jaw and the first molars on each side of the lower jaw are especially modified to form the carnassial teeth, which are effective at slicing through the flesh
- sharply pointed premolars and molars.

In addition, the jaw articulation is tight, with larger chewing muscles than the herbivores (where there is more of a rotational chewing movement).

The gut of a carnivore is not significantly different from the human gut, but there are other adaptations shown. The carnivore is a hunter and depends on good vision and speed in order to catch live prey. Comparison of the skull of a dog or cat with that of a herbivore reveals the difference in the position of the eyes. The carnivore has eyes at the front, positioned quite close together to give stereoscopic vision, thus enabling distances to be judged accurately. The herbivore's eyes are on the side of its head and give a more all-round view, detecting movement of predators more easily. Some carnivores have good night vision, which enables them to hunt at night and surprise their prey.

Other adaptations that may be seen in the carnivores include:
- the ability to run fast in order to chase their prey
- camouflage (coat colour or markings) so that they can stalk their victims without being seen
- powerful claws on their limbs to hold their prey.

Members of the Order Carnivora include leopards, cheetahs and the domestic cat which usually hunt their prey singly; others, such as the wolves and hyenas, hunt in social groups called packs. Lions may hunt singly or in packs. Not all carnivores hunt large prey that have to be chased and caught. Some, like

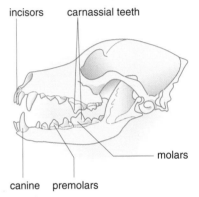

Figure 6.5 Skull of a domestic dog showing features of carnivore dentition.

badgers, eat earthworms which they collect from the surface of the soil. Otters hunt and catch fish underwater.

Saprobiontic nutrition

So far in this chapter we have been considering in detail the type of heterotrophic nutrition shown by most animals, namely **holozoic** nutrition. **Saprobiontic**, or saprotrophic, organisms obtain their nutrition from dead or decaying organic remains of plants and animals. They are **consumers** in **detritus food chains** and are often referred to as **decomposers**. Saprobionts secrete enzymes onto the organic matter and absorb the soluble products of this **extracellular** digestion. Any substances released by this digestion which are not taken up by the saprobionts are made available for uptake by plants, so contributing to the circulation of nutrients. Many bacteria and fungi are saprobiontic.

Rhizopus – a saprobiontic organism

Fungi are eukaryotic organisms that lack chlorophyll and are therefore non-photosynthetic. Their hyphal walls contain chitin. *Rhizopus stolonifera* is a member of the Zygomycota, which are fungi lacking cross-walls in their hyphae. *Rhizopus* (Figure 6.6) is commonly found growing on damp wholemeal bread, where it can be seen as cotton-like white threads on the surface. After a few days, masses of tiny black sporangia appear, resembling pin heads and giving the fungus its common name 'pin mould'. Closer observation, using a microscope, reveals that the fungus consists of a **mycelium** made up of much-branched **hyphae**, which are **aseptate** (lacking cross-walls). Aerial hyphae, called **stolons**, spread over the food substrate and produce tufts of branched hyphae, or **rhizoids**, where they touch down. The rhizoids penetrate the substrate, secreting enzymes which digest the complex substances in the food. The soluble products of this digestion are absorbed by the rhizoids and either used in metabolic activities or stored. Three main groups of enzymes are secreted as in other types of heterotrophic organisms: carbohydrases, proteases and lipases.

Parasitic nutrition

A **parasite** is an organism which lives in close association with another living organism, the **host**. In this relationship, the parasite is dependent on the host for its food and usually causes the host some degree of harm. **Ectoparasites** live on the outside of their hosts while **endoparasites** live inside their host. Most endoparasites, such as gut parasites, spend their entire lives within their host, but some ectoparasites, such as the flea, only become attached during feeding. Most parasites are highly adapted to their particular mode of life.

Taenia – a parasitic organism

Taenia solium, the pork **tapeworm** (Figure 6.7), is an endoparasite with two hosts. Its primary host is man, and the adult stage is found attached to the wall of the small intestine of an infected individual. The secondary host is the pig, in which the larval stage develops. The adult stage consists of a flattened, ribbon-

(a)

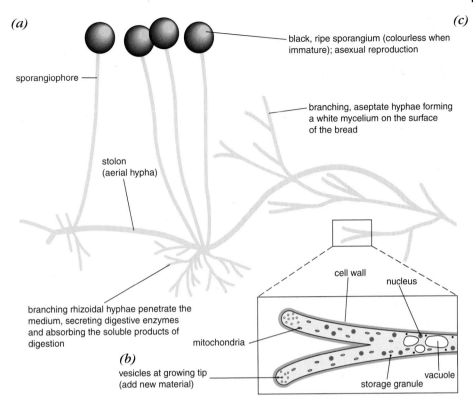

sporangiophore —

black, ripe sporangium (colourless when immature); asexual reproduction

branching, aseptate hyphae forming a white mycelium on the surface of the bread

stolon (aerial hypha)

branching rhizoidal hyphae penetrate the medium, secreting digestive enzymes and absorbing the soluble products of digestion

mitochondria

cell wall

nucleus

(b)

vesicles at growing tip (add new material)

vacuole

storage granule

(c)

head (scolex)

hooks

sucker —

zone of proliferation

Figure 6.6 (a) Rhizopus *on bread, showing sporangia; (b) diagram of* Rhizopus *showing mycelial structure; (c) photomicrograph of sporangia of* Rhizopus

like body, made up of a large number of 'segments' called **proglottides** (sing. **proglottis**). The tapeworm can be up to 3.5 m long and is approximately 6 mm wide and 1.5 mm thick. At its anterior end, it has a tiny knob known as the **scolex** which has a double row of hooks and four suckers. Just behind the scolex is a region called the **proliferation zone** where new proglottides form. The rest of the organism consists of proglottides containing both male and female reproductive structures. At the posterior end, after self-fertilisation has occurred, these proglottides contain a greatly enlarged uterus full of fertilised eggs, the rest of the structures having been reabsorbed. These proglottides become detached from the organism and pass out with the faeces of the host. If ingested by the secondary host, a pig, the life cycle proceeds to the next stage, giving rise to a larval stage in muscular tissue.

Taenia has no mouth or alimentary canal, as it absorbs the digested food of its host all over its body surface. Simple, soluble products of digestion of the host's food will be present in the small intestine, so the tapeworm can absorb the food it requires and has no need of a digestive system of its own, nor does it need to secrete digestive enzymes. It has a thick outer covering, known as the **tegument**, consisting of protein and chitin, which protects it from the digestive enzymes of its host. Other adaptations to its mode of life are:
- its ability to live in the low oxygen concentrations present in the human gut
- a reduction in the nervous system and a lack of sense organs
- the possession of suckers and hooks for attachment to the gut wall of the host
- the production of very large numbers of offspring.

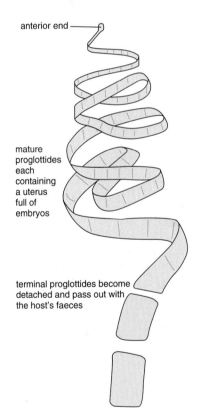

anterior end —

mature proglottides each containing a uterus full of embryos

terminal proglottides become detached and pass out with the host's faeces

Figure 6.7 Taenia, *adult stage with detail of anterior region*

If the human host is healthy, little damage is caused by the tapeworm apart from depriving the host of some of its food. The long thin shape of the tapeworm does not block the small intestine. In children and adults already debilitated by other diseases, it can cause problems. Individuals may become less resistant to other diseases and suffer from abdominal pain, vomiting, constipation and loss of appetite.

Mutualism

The term **mutualism** describes a close association between two organisms in which both contribute and both benefit. This type of association can be illustrated by reference to the relationship between the bacterium *Rhizobium* and members of the flowering plant family Papilionaceae.

Rhizobium is a nitrogen-fixing bacterium. In nitrogen fixation, hydrogen ions from carbohydrates, such as glucose, are combined with nitrogen to form ammonia. The ammonia then combines with glutamate to form the amino acid glutamine, from which other amino acids can be synthesised. The fixation reaction takes place in anaerobic conditions in the cytoplasm of the bacterial cells, catalysed by the enzyme **nitrogenase** (Figure 6.8).

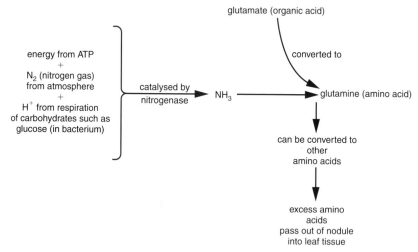

Figure 6.8 Stages in nitrogen fixation – amino acids are synthesised by the bacteria in the nodules

Rhizobium bacteria are present in the soil. Those near the roots of leguminous plants, such as clover, peas or beans, are attracted towards the roots and are able to penetrate the root hair cells. This attraction has been found to occur in response to a hormone secreted by the roots of the plant. The bacteria also secrete substances which cause the root hair cells to bend around, possibly encouraging penetration. Once inside the root, the bacteria move to the cortex and stimulate the production of auxins and cytokinins by the root tissue. This causes cell division to occur and a nodule of tissue is formed in the cortex, composed of cells containing large numbers of the bacteria. The bacteria become Y-shaped and have a banded appearance. These forms of the bacteria are called **bacteroids**. The bacteroids are able to fix nitrogen in the nodules, where the anaerobic environment is suitable for the efficient functioning of their nitrogenase enzymes. Oxygen molecules are absorbed by a special pigment, **leghaemoglobin**, which surrounds the bacteroids and gives a pinkish colour to the tissues of the nodule. Figure 7.14 on page 148 shows root nodules in the bean, *Phaseolus multiflorus*, and a TS through a nodule is shown in Figure 6.9.

The bacteria benefit from this association by obtaining their supplies of carbohydrate from the photosynthetic activities of the plant. The plant benefits by receiving a supply of ammonia from the bacteria. This enables the leguminous plants to grow in nitrogen-deficient soils.

The relationship between a cow and the bacteria in its rumen is also an example of mutualism. The bacteria benefit from being supplied with a food source, cellulose, and with a suitable environment for growth. The cow benefits from the breakdown of the cellulose into fatty acids, which can be absorbed into the blood through the rumen wall. The bacteria also synthesise proteins and some B group vitamins which may further benefit the cow.

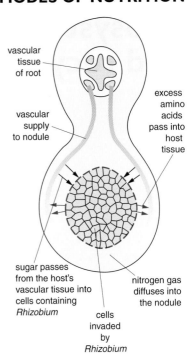

vascular tissue of root

vascular supply to nodule

excess amino acids pass into host tissue

sugar passes from the host's vascular tissue into cells containing *Rhizobium*

cells invaded by *Rhizobium*

nitrogen gas diffuses into the nodule

Figure 6.9 TS through a root nodule

7 Ecosystems, energy flow and recycling of nutrients

Ecosystems

Figure 7.1 A deciduous woodland (in the UK) in April – an example of a habitat

The term **biosphere** encompasses the part of the Earth and its atmosphere inhabited by living organisms. Within the biosphere are many different ecosystems. These represent a complex series of interrelationships between the rock, soil, water, air and living organisms.

An **ecosystem** was first defined in 1935 by Sir Arthur Tansley as 'the living world and its habitat'. This definition includes both the organisms within the ecosystem (the **biotic** component) and the physical and chemical factors (the **abiotic** component) which influence them. Different ecosystems may have more or less clearly defined boundaries, but they also merge into one another, so it may not always be easy to see the boundaries. For example, there is a gradual transition from a freshwater ecosystem, such as a pond, through the surrounding marshy ground to a field.

An ecosystem will contain a number of different **habitats**. A habitat is the place within an ecosystem where an organism lives. The pond shown in Figure 7.2 is a habitat for a whole range of different species, including pond snails and phytoplankton. Many organisms occupy one particular part of the total habitat, for example, the pea mussels living in the mud at the bottom of the pond. This location within the overall habitat is sometimes referred to as a **microhabitat**.

An ecosystem thus comprises several habitats with their **communities** of organisms, and includes both biotic and abiotic components. A useful way of thinking of an ecosystem is as a network of habitats interlinked by the flow of energy and nutrients. Abiotic components are the non-living factors that influence the distribution of organisms within an ecosystem. These components include physical factors (such as light intensity, temperature, rainfall, wind and water currents) and chemical factors (including all the elements and compounds in the ecosystem). Biotic components also affect the populations of

> **DEFINITION**
>
> The term **biosphere** refers to the part of the Earth which is inhabited by living organisms. An **ecosystem** consists of communities of organisms which interact with each other and their physical and chemical environment. The place where an organism lives is known as the **habitat**, such as a pond or rocky shore.

> **DEFINITION**
>
> A **community** is a number of populations of organisms living in a habitat.

138

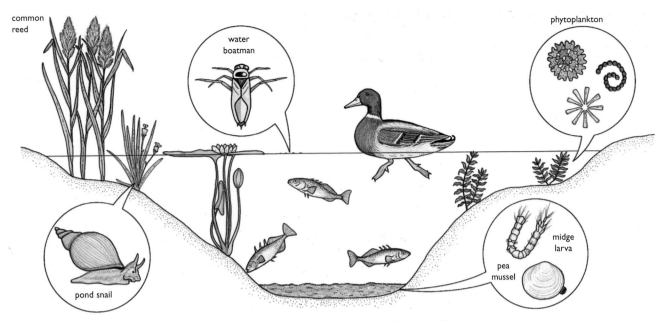

Figure 7.2 A freshwater ecosystem. Organisms in the inserts are shown enlarged, but not to the same scale.

organisms in an ecosystem; these include the availability of food, competition and predation.

Ecosystems are complex associations of plants, animals and microorganisms which interact with each other and with their non-living environment. The flow of energy and flow of nutrients within ecosystems are determined by the organisms living there. Through photosynthesis, green plants, the **producers**, incorporate about 1 per cent of the sun's energy falling on them into organic compounds. The plants may then be eaten by herbivores (**primary consumers**), which in turn may be eaten by carnivores (**secondary consumers**). This type of sequence forms a **food chain** and the different levels within it are called **trophic levels.**

Energy flow

All organisms in an ecosystem depend on an adequate supply of energy for their survival. Energy from the sun, trapped by photosynthesis, provides the source of energy for all living organisms. Carbohydrates, which are produced in photosynthesis, are used both as building blocks for the growth of green plants and as an energy source for the plant. Carbohydrate is respired in plant cells to provide adenosine triphosphate (ATP) for reactions, such as protein synthesis.

Heterotrophs cannot utilise carbon dioxide by photosynthesis and therefore depend on obtaining ready made organic compounds from their food. In heterotrophs, these organic compounds serve the same functions as they do in autotrophs, that is, they provide building blocks for the synthesis of new cell compounds for growth and provide substrates for the production of ATP by respiration.

ECOSYSTEMS, ENERGY FLOW AND RECYCLING OF NUTRIENTS

The roles of producers, consumers and decomposers

We have already stated that energy from the sun enters an ecosystem through organic compounds produced, in photosynthesis, by green plants.

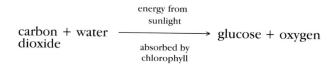

$$\text{carbon dioxide} + \text{water} \xrightarrow[\substack{\text{absorbed by} \\ \text{chlorophyll}}]{\substack{\text{energy from} \\ \text{sunlight}}} \text{glucose} + \text{oxygen}$$

Figure 7.3 Green plants use light energy to synthesise organic substances from carbon dioxide and water, in the process of photosynthesis. These organic substances are used for growth and as an energy source. Heterotrophs use the organic compounds from the plant for growth and for energy.

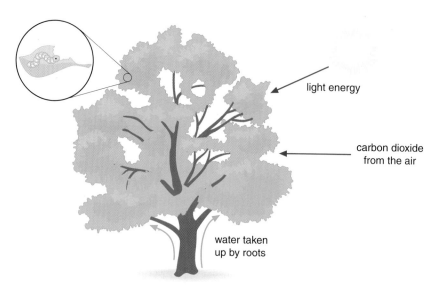

Green plants are described, in this context, as **primary producers**. The Earth's surface receives between 0 and 5 joules of solar energy per square metre every minute. However, only a small percentage of this energy is captured by chlorophyll and converted into chemical energy of newly synthesised organic compounds. Some of the light striking a leaf is reflected, or passes straight through, and about half is made up of wavelengths which are not used in photosynthesis. There is also some wastage of energy due to biochemical inefficiency of the reactions of photosynthesis.

Transfer of energy through food chains and food webs

Figure 7.4 shows an example of a **food chain**, that is, a sequence of organisms feeding on other organisms.

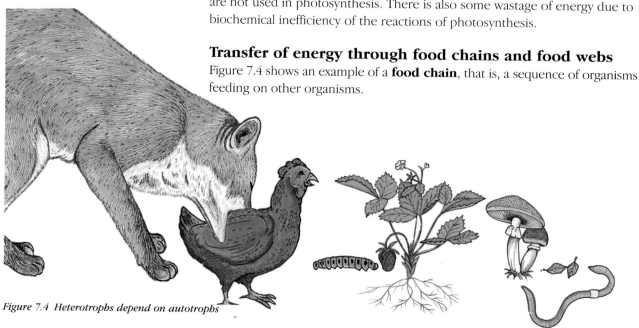

Figure 7.4 Heterotrophs depend on autotrophs

ECOSYSTEMS, ENERGY FLOW AND RECYCLING OF NUTRIENTS

One sequence shown is:

plant → caterpillar → chicken → fox

Each organism within the chain occupies a particular **trophic level**. The food chain 'plant to fox' shown above has four trophic levels and can be written in general terms as:

primary producer → herbivore → first carnivore → second carnivore

There are very large numbers of different food chains in the range of ecosystems on Earth, but it is rare to find chains with more than five trophic levels. Green plants fix only about 1 per cent of the sun's energy that falls on their leaves, and successive members of a food chain incorporate into their own biomass about 10 per cent of the energy available in the organisms they consume. The loss of energy at each trophic level is so great that very little of the original energy remains in the chain after it has been incorporated successively into the biomass of organisms at four trophic levels.

If you look carefully at Figure 7.4, you will see other relationships between the organisms. As examples: dead plant leaves provide a food source for earthworms; fungi are decomposers and also obtain nutrients from dead leaves; worms and caterpillars are eaten by both birds and foxes. It is clear that in this diagram there are several food chains, and one organism may feed at more than one trophic level – foxes can be first-level or second-level carnivores, or even herbivores, as they are partial to fruit. There is a complex series of feeding relationships between organisms in an ecosystem. This set of relationships is referred to as a **food web**. Figure 7.5 shows a generalised food web based on Figure 7.4.

> **DEFINITION**
> A **food web** illustrates the feeding relationship between different organisms in an ecosystem. Food webs consists of a number of interlinked food chains.

Constructing a food web involves detailed field and laboratory work. The main approaches include observation of predator–prey relationships, laboratory food preference experiments, analysis of gut contents and the use of radioactive isotopes such as ^{32}phosphorus (^{32}P). Plants can be labelled with ^{32}P and the passage of this isotope is then followed in nearby animal populations.

An example of a food web is illustrated in Figure 7.6. This shows some of the complex feeding relationships between organisms on a rocky shore.

> **QUESTION**
> Identify the primary producers, herbivores, and first and second carnivores in the food web illustrated in Figure 7.6. How many trophic levels are present?

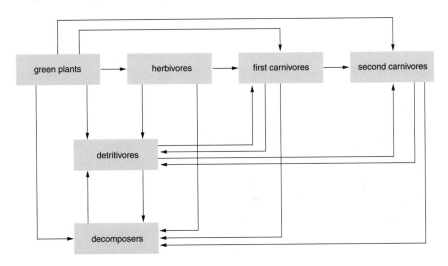

Figure 7.5 A generalised food web (the arrows show the direction of energy flow)

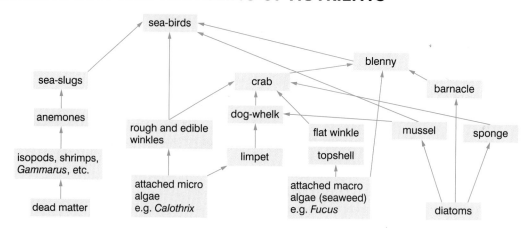

Figure 7.6 Part of a food web on a rocky shore

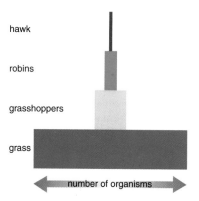

Figure 7.7 A pyramid of number is a diagrammatic way of representing the numbers of organisms in different levels of a food chain

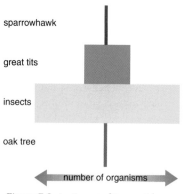

Figure 7.8 An inverted pyramid of number

Pyramids of number, biomass and energy

In an ecosystem, there are usually far more organisms at lower trophic levels than at higher trophic levels. Similarly, the biomass of primary producers present in an ecosystem is much greater than that of the herbivores, and successive trophic levels have a progressively smaller biomass.

We can represent these relationships as diagrams, known as **pyramids**. An example of a **pyramid of number** is shown in Figure 7.7. The width of each block is proportional to the numbers of organisms present at each trophic level.

One disadvantage of representing a food chain in this way is that it does not take into account the biomass of each trophic level. For example, one rose bush might support a very large population of aphids, so if this was represented diagrammatically, we would have an inverted pyramid of number (such as that illustrated in Figure 7.8).

A **pyramid of biomass** shows the mass of material at each trophic level. This usually results in an upright pyramid – the biomass of a rose bush would be much greater than the biomass of the aphids it supports. However, there are situations where an inverted pyramid of biomass can be obtained. Pyramids of biomass show the standing crop, that is, the biomass at one particular time, and do not take into account the fact that the biomass at each trophic level may vary over a period of time. For example, in January, the biomass of animal plankton in the Channel is greater than that of the plant plankton (the primary producers). However, over the whole year, the total biomass of primary producers far exceeds that of the consumers in this ecosystem.

A **pyramid of energy** (Figure 7.9) gives a more accurate representation of the transfer of material from one trophic level to the next. Rather than plotting numbers, or biomass, of organisms at each trophic level, we plot the productivity for each level in the ecosystem. Productivity is a measure of the energy content of each level and can be obtained by converting the mass of new organic material produced per unit area per year into an equivalent energy value. This energy value for each trophic level is expressed in units of $kJ\ m^{-2}\ yr^{-1}$.

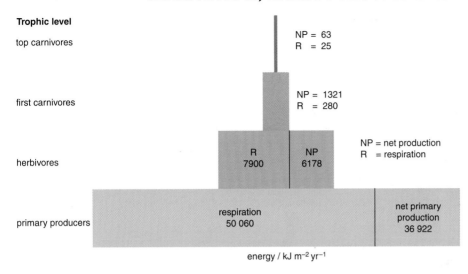

Trophic level

top carnivores — NP = 63, R = 25

first carnivores — NP = 1321, R = 280

herbivores — R 7900, NP 6178

NP = net production
R = respiration

primary producers — respiration 50 060, net primary production 36 922

energy / kJ m^{-2} yr^{-1}

Figure 7.9 A pyramid of energy shows the energy content at each trophic level

Productivity

All the living organisms within a unit area constitute the standing crop, or **biomass**. Biomass is defined as the mass of organisms per unit area of ground and may be expressed in units such as kilograms per hectare (kg ha^{-1}). Biomass may also be converted to an equivalent energy value and in this case is expressed in terms of units of energy per unit area, such as kilojoules per hectare (kJ ha^{-1}).

The **primary productivity** of an ecosystem refers to the rate at which biomass is produced per unit area, by green plants. Primary productivity may be expressed either in units of mass of dry matter produced, such as kilograms, or in units of energy, such as kilojoules. Since we consider productivity in relation to area and time, the units for the rate of accumulation of biomass are kilojoules per unit area per year, for example, kilograms per hectare per year (ka ha^{-1} yr^{-1}). In terms of energy, this could be expressed as kilojoules per unit area per year (for example kJ ha^{-1} yr^{-1}).

The total amount of energy captured by green plants in the process of photosynthesis is referred to as the **gross primary production** (**GPP**). This is expressed in terms of units of energy per unit area per year, usually kJ per metre squared per year (kJ m^{-2} yr^{-1}). The plant uses some of the organic materials produced in photosynthesis as substrates for respiration, so some of the GPP will be lost. The GPP used by the plant for respiration is ultimately lost as heat energy. The rate at which organic compounds are used in this way is referred to as **plant respiration** (**R**) and is also measured in kJ m^{-2} yr^{-1}.

The difference between gross primary production (GPP) and losses due to respiration (R) is known as **net primary production** (**NPP**). The net primary production is important because it represents the biomass that is available for consumption by heterotrophic organisms.

ECOSYSTEMS, ENERGY FLOW AND RECYCLING OF NUTRIENTS

We can summarise the relationship between GPP, NPP and R by the equation:

$$GPP = NPP + R$$

This relationship, and the fates of solar energy falling on a leaf, are shown in Figure 7.10.

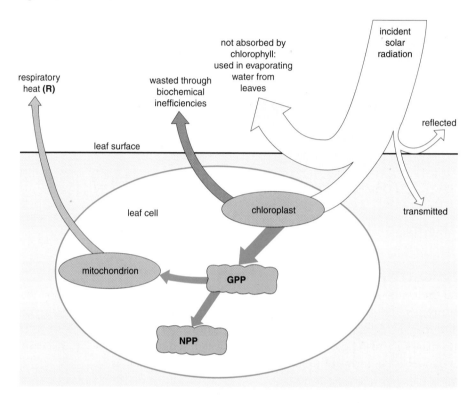

Figure 7.10 Fates of solar energy falling on a leaf

It is estimated that the global terrestrial net primary production is 110 to 120 $\times$ 10^9 tonnes dry mass per year, and in the sea, 50 to 60 $\times$ 10^9 tonnes per year. Since only the NPP is available for consumers to eat, NPP values are often used to compare the **productivity** of ecosystems. Productivity varies widely in different ecosystems, as shown in Table 7.1.

Table 7.1 *Mean values for net primary production (NPP) in a range of ecosystems*

Ecosystem	Mean NPP / kJ m^{-2} yr^{-1}
extreme desert	260
open ocean	4700
temperate grasslands	15 000
temperate deciduous forest	26 000
intensive agriculture	30 000
tropical forest	40 000

QUESTION

Apart from the availability of light energy, what factors will influence the primary productivity of different ecosystems?

What happens to the NPP in an ecosystem? Some NPP will remain stored within plants and increase plant biomass, and some may be eaten by herbivores. This energy input to herbivores is referred to as **herbivore consumption**. In a grass field, about 40 per cent of the annual production may be eaten by cows, but in a forest only about 2 per cent of the NPP passes to herbivores. Herbivores may themselves be preyed upon by predators, the first carnivores. The energy input to these animals is termed **carnivore consumption**. A final part of the NPP, contained in, for example, dead leaves and flowers, reaches the ground where it forms litter or **detritus**. This provides a source of food for a variety of organisms. Some of these are earthworms, termed **detritivores**; others are soil fungi and bacteria. These soil microorganisms are collectively known as **decomposers** and are the ultimate consumers of all dead organic matter in an ecosystem.

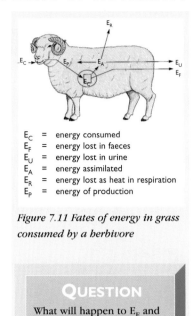

E_C = energy consumed
E_F = energy lost in faeces
E_U = energy lost in urine
E_A = energy assimilated
E_R = energy lost as heat in respiration
E_P = energy of production

Figure 7.11 Fates of energy in grass consumed by a herbivore

QUESTION

What will happen to E_F and E_U shown in Figure 7.11?

We have already stated that successive members of a food chain incorporate into their biomass only about 10 per cent of the energy available in the food they eat. What happens to the remaining 90 per cent?

Let us consider the possible fates of the energy available in grass eaten by a herbivore, such as a sheep. Some of the grass will remain undigested and will be egested as faeces, so this represents one source of energy loss. The grass which is digested will be absorbed from the gut and assimilated by the herbivore. There are three possible fates of the energy contained in this assimilated grass:
1 It can be used in respiration and will be lost as heat
2 It can contribute to an increase in the biomass of the herbivore, termed **energy of production**
3 A small amount will be lost in urine.

These possible fates are illustrated in Figure 7.11.

We can write an equation which summarises the fate of all the energy consumed by the herbivore:

$$E_C = E_P + E_F + E_U + E_R$$

This equation accounts for all of the energy entering the herbivore. The energy of production, E_P, will be available as energy of consumption for a carnivore, and the sequence is repeated at the next trophic level. Only about 10 per cent of the energy entering one trophic level (the energy of consumption) is available for consumption by the next trophic level.

Recycling of nutrients

We have seen that a continuous supply of energy is essential for all living organisms. Organisms also require a range of chemical substances, including water, mineral ions and organic compounds. The original supply of energy is the sun, but there are only fixed amounts of chemical substances available on Earth. The chemicals needed to build the tissues of living organisms are used and re-used repeatedly. These substances move in cycles from the soil, water or the atmosphere, into plants and animals and back again. Cycles that operate in this way are known as **biogeochemical cycles** and include the carbon, sulphur and nitrogen cycles. The **water** (**hydrological**) **cycle** is shown in Fig 7.12.

DEFINITION

Biomass refers to the mass of living material per unit area, or per unit volume in an aquatic habitat. **Gross primary production** (GPP) is the rate of formation of organic material by green plants. Some of this material will be used by the plants in the process of respiration and ultimately lost as heat energy. The remainder is referred to as the **net primary production** (NPP).

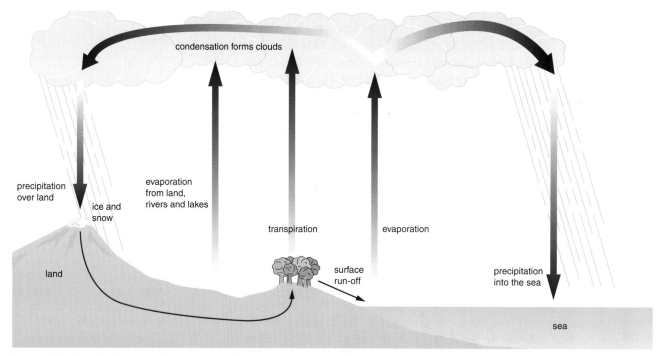

Figure 7.12 The hydrological cycle (or water cycle) consists of a series of natural processes by which water evaporates and forms clouds, falls as rain or snow, and eventually returns to oceans via streams and rivers

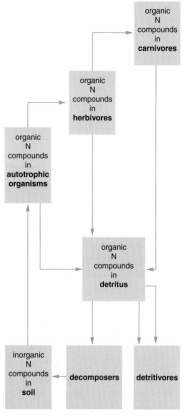

Figure 7.13 The biological part of the nitrogen cycle

In its simplest form, the water cycle is a sequence of natural processes in which water **evaporates** into the atmosphere, cools and **condenses** to form clouds, then falls back to the surface of the Earth in the form of rain or snow. The term **precipitation** refers to the ways in which water falls back to the Earth as rain or snow, and includes hail and sleet. Much of the water taken up by plant roots is also lost by evaporation to the atmosphere, mainly through stomata on the leaves. This process is known as **transpiration**.

The nitrogen cycle

Nitrogen is required by all living organisms, because it is a component of many substances, including nucleic acids and proteins. Plants and animals obtain their nitrogen in different ways. The source of nitrogen for green plants is in the form of inorganic ions, either nitrate (NO_3^-) or the ammonium ion (NH_4^+) present in soil or water. Heterotrophic organisms obtain their nitrogen only from organic substances.

In a typical green plant, nitrate ions taken up by the roots are reduced to ammonium ions in the cells and are then incorporated into organic compounds to form, for example, amino acids. These organic compounds then provide the source of nitrogen for detritivores, herbivores and, subsequently, carnivores.

Organic nitrogen-containing compounds in dead tissues, faeces and urine can be converted back into inorganic nitrate by the action of saprophytic bacteria and fungi, referred to as **decomposers**. We can illustrate the biological part of the nitrogen cycle in a simplified form, as in Figure 7.13.

ECOSYSTEMS, ENERGY FLOW AND RECYCLING OF NUTRIENTS

Bacterial and fungal decomposers break down organic nitrogen compounds in dead and decaying organic matter to release ammonium ions (NH_4^+). If no oxygen is available, or if the soil is waterlogged, cold or acidic, the process stops here and only ammonium ions are available to plants. In 'good' soil conditions, ammonium ions are oxidised by other soil bacteria (such as *Nitrosomonas*) into nitrites (NO_2^-). Another group of soil bacteria (such as *Nitrobacter*) further oxidise nitrite ions into nitrate ions (NO_3^-). The overall conversion of ammonium ions to nitrate ions is termed **nitrification**.

$$\underset{\text{ammonium}}{NH_4^+} \quad \xrightarrow{\textit{Nitrosomonas}} \quad \underset{\text{nitrite}}{NO_2^-} \quad \xrightarrow{\textit{Nitrobacter}} \quad \underset{\text{nitrate}}{NO_3^-}$$

Nitrate ions produced in this way are available for uptake by plant roots. However, nitrate ions are very soluble in water and, unlike other ions present in the soil, do not bind tightly to soil particles. As a consequence, nitrate ions are easily washed out of soil after heavy rain, in a process called **leaching**.

Another way in which nitrate ions can be lost from soil is through the process of **denitrification**. Under anaerobic conditions, such as when soils are waterlogged, denitrifying bacteria convert nitrate to nitrite and then to nitrogen gas, which escapes into the atmosphere. Denitrifying bacteria include *Pseudomonas denitrificans* and *Thiobacillus denitrificans*.

Fertilisers containing nitrogen in the form of ammonium nitrate or ammonium sulphate may be added to soils to increase the production of agricultural crops. These fertilisers are manufactured industrially by the Haber–Bosch process in which nitrogen and hydrogen react together at high temperatures and pressures, in the presence of an iron catalyst, to form ammonia. However, it has been estimated that crops take up only about 50 to 80 per cent of the applied nitrogen; the rest is lost in the processes of leaching and denitrification. Leaching of nitrate fertilisers may lead to **eutrophication**; which is described in Chapter 9.

There are several ways in which nitrates can be added to the soil. They can be added as nitrogen-containing fertilisers in managed ecosystems. Small amounts of inorganic nitrogen compounds are also formed by the action of lightning in the atmosphere. This produces oxides of nitrogen which combine with rainwater to form nitrate ions.

Nitrogen-fixing organisms, which live in soil, are able to reduce nitrogen gas to ammonia. This is a biological version of the Haber–Bosch process. Unlike this chemical nitrogen fixation, which requires temperatures of 300 to 500 °C, high pressures and an iron catalyst, biological nitrogen fixation is much more efficient, occurring at low temperatures and at atmospheric pressure. Biological nitrogen fixation is catalysed by nitrogenase, a complex enzyme containing iron and molybdenum.

$$\underset{\text{nitrogen}}{N_2} + \underset{\text{hydrogen}}{3H_2} \xrightarrow{\text{nitrogenase}} \underset{\text{ammonia}}{2NH_3}$$

> **DEFINITION**
>
> The **Haber–Bosch** process is an industrial method for the production of ammonia, which can then be used to make nitrogen fertilisers. In this process, nitrogen and hydrogen react together at high pressure and high temperature to produce ammonia.

ECOSYSTEMS, ENERGY FLOW AND RECYCLING OF NUTRIENTS

Figure 7.14 Nodules on the roots of the runner bean, Phaseolus multiflorus

QUESTIONS

How do the following human activities affect the recycling of carbon or of nitrogen:
• burning fuels
• deforestation
• harvesting of agricultural crops
• intensive rearing of animals
• use of nitrogen fertilisers?
List any other human activities which are likely to disrupt the carbon and nitrogen cycles. [*You may find help with your answers in Chapters 8 and 9.*]

Only certain bacteria and cyanobacteria can fix nitrogen gas in this way. Some of these organisms, such as *Azotobacter vinelandii*, are free-living in the soil. One genus of nitrogen-fixing bacteria, *Rhizobium*, forms a mutualistic (symbiotic) relationship with legumes (plants such as peas, beans and clover). These plants develop swellings on their roots, called **root nodules**, containing *Rhizobium* (Figure 7.14). It is a symbiotic relationship because the bacteria receive carbohydrates and ATP from the plant, which obtains fixed nitrogen, in the form of ammonia, in return. Ammonia is then incorporated into organic compounds to synthesise amino acids.

One of the long-term aims of gene technology is to incorporate genes for nitrogen fixation into non-leguminous plants. This could have considerable environmental and economic benefits, avoiding the need for expensive inorganic fertilisers, only about half of which are actually taken up by plants. Details of how human activities can disrupt the nitrogen cycle can be found in Chapter 9.

The complete nitrogen cycle, incorporating both biological and chemical elements, is illustrated in Figure 7.15.

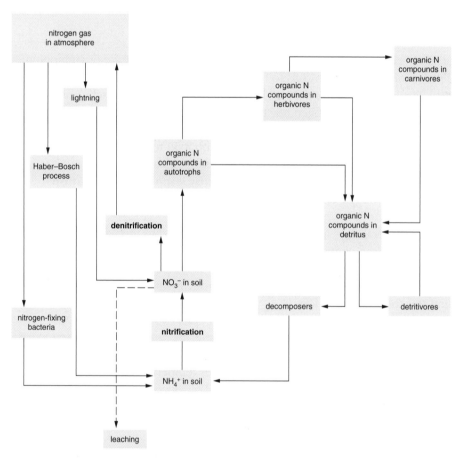

Figure 7.15 The complete nitrogen cycle

The carbon cycle

Carbon dioxide (CO_2) present in the atmosphere or dissolved in water, is taken up by photosynthetic organisms, that is, the primary producers in a food chain. Carbon dioxide is used to produce organic substances, such as glucose, in the process of **photosynthesis**. Primary producers may be eaten by herbivores, carnivores may eat herbivores, and so the organic substances are passed along a food chain. All living organisms respire. Some of the organic substances made by the primary producers will be used as substrates for **respiration**. Consumers and decomposers also use organic substances as substrates for respiration. Respiration produces carbon dioxide and, as a result, carbon dioxide is returned to the atmosphere. Remains of dead organisms, leaves that fall from trees in the autumn, and excreted substances decay and are broken down by microorganisms. These too release carbon dioxide produced in respiration. These processes are illustrated in Figure 7.16.

Fossil fuels, such as coal and oil, are 'sinks' for carbon dioxide as they were formed from the incomplete breakdown of plant material many millions of years ago. Coal and oil are referred to as sinks for carbon dioxide as they contain carbon, which has effectively been trapped and can only be released by combustion. Combustion (burning) of fossil fuels produces carbon dioxide, which is thus returned to the atmosphere.

Carbon dioxide in water can form carbonate ions (CO_3^{2-}) and hydrogencarbonate ions (HCO_3^-). These ions contribute to the formation of carbonate rocks, such as limestone. The processes of weathering through, for example, the action of acid rain on carbonate rocks, and volcanic activity can return carbon dioxide to the atmosphere.

Human activities, such as burning of fossil fuels and deforestation, disrupt the carbon cycle and are believed to be responsible for the gradual increase in atmospheric carbon dioxide concentrations. This topic is described in more detail in Chapter 9.

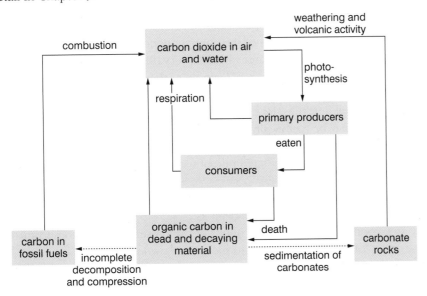

Figure 7.16 The carbon cycle

8 Energy resources

Figure 8.1 Yak dung for fuel – in the high grasslands of western China (at around 3500 m, close to Tibet), there is not enough woody growth (as trees or shrubs) to supply the fuel needs of local people. Yak and cattle dung becomes a valuable resource. The lower photograph shows a 'fuel store' inside the tent of nomadic herdsmen. Dung is also valuable for people living in desertified areas, where woody cover has become too sparse to supply fuel requirements

Energy resources and sustainability

When early humans began to live in settled communities, they developed means of using external sources of energy as part of their way of life. Today, the main categories of energy use are essentially similar: to provide low temperature heat for human comfort and to cook food, to generate heat at a temperature high enough to give light and to work materials in a way that has led to a vast range of industrial processes. We also harness energy to provide a force that allows movement and mechanical work to be done, which by now has expanded to include many different forms of transport.

For centuries, traditional energy sources have included wood or dung mixed with straw for fuel, and animals, wind and water for power. A change in emphasis for fuel sources came with the greater requirements linked to the Industrial Revolution, and these demands have continued to increase up to the present day. Alongside this change in emphasis there has been heavy exploitation of fossil fuels, for generation of electricity as an almost universal source of energy (for heat, light and movement) together with the use of oil for the internal combustion engine.

Over the past 40 or so years, there has been a growing concern over the consequences of using such high levels of energy in our societies on a global scale. Since the late 1960s, there has been increasing recognition of the **environmental impact** of energy use, particularly of fossil fuels. The harmful effects relate mainly to carbon dioxide output, the enhanced greenhouse effect and how far this affects global warming, and also to acid rain and to oil pollution of the seas. These pollution effects are considered in more detail in Chapter 9.

The other major concern has been over the **sustainability** of our energy resources. There were worries, particularly in the 1970s, that reserves of fossil fuels would not last the lifetime of the younger generation at that time, if use continued at the same or a greater rate. It is very difficult to make realistic and reliable estimates, but the balance of opinion at the start of the 21st century is that more fossil fuel reserves do exist and are sufficient to sustain global energy requirements well into the 21st century. However, as we dig deeper (literally), the fossil fuels may be technically more difficult to exploit and hence more expensive. Ultimately, however, such reserves must be finite since the rate of formation lags far behind predicted rate of use in the 21st century.

A statement about **sustainable development** was made in the Brundtland Report in 1987. The definition was given as:
> '… development that meets the needs of the present without compromising the ability of future generations to meet their own needs.'

If we interpret this in terms of sustainability of energy resources, the important point is that the energy resource is not depleted or used up and can be replenished (renewed) naturally within a reasonable time-scale.

A shift towards greater use of renewable energy sources (see pages 153, 155) would go a long way in helping to sustain our global energy sources. Management of this will only be achieved by firm incentives at a national level to encourage use of energy from renewable sources, together with further development of the technology required to harness the energy in different ways. The alternative would be to make drastic reductions in the high demands for energy that have become a part of our 21st century lifestyle.

Use of fossil fuels

Fossil fuels (**coal**, **oil** and natural gas) are derived from biomass that was living millions of years ago. On a global scale, these three fossil fuels provide the major source of energy, probably around 90% of total energy consumption. The figure is higher in Europe and other industrialised countries, whereas in developing countries there is a greater dependence on traditional fuels such as firewood and charcoal. Global trends for use of fossil fuels are illustrated in Figure 8.2. These charts emphasise the increasing dependence on fossil fuels but they also illustrate a gradual reduction in the use of coal and an increase in the use of natural gas.

The ways in which these fossil fuels are used are summarised at the start of this chapter. In addition to their use as an energy source (including provision of heat, light, power and fuel oils for transport), both coal and oil provide raw materials that are used industrially for the manufacture of a wide range of products including plastics, dyes, drugs and other chemicals.

(a)

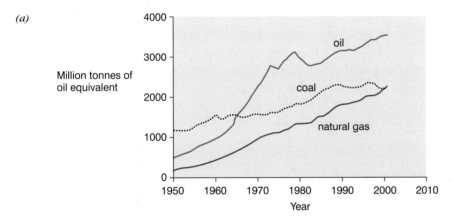

(b)

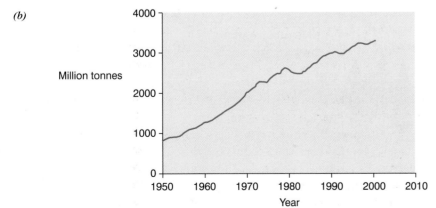

Figure 8.2 (a) World consumption of fossil fuels (coal, oil and natural gas) from 1950 through to 2001; (b) World emissions of carbon from burning of fossil fuels, from 1950 through to 2001

You need to have an understanding of the biological origin of fossil fuels, partly because of their position in the carbon cycle (Chapter 7, page 149) and also to appreciate why there is an increasing pressure for using alternative energy sources. Reserves of fossil fuels are limited because the rate at which they are formed is very much slower than the rate at which they are being used.

Fossil fuels are relatively cheap and easy to obtain, transport and store. The chief disadvantage of burning fossil fuels lies in their impact on the environment, because of the pollution caused by various emissions from industry and from motor vehicles. The effects of acid rain, and the build-up of carbon dioxide with possible long-term effects on global climate, are largely attributed to the use of fossil fuels (see Chapter 9). Coal is considered to be the dirtiest of the three, causing most environmental damage, whereas natural gas is the cleanest.

> ### QUESTION
>
> **Why do you think use of fossil fuels has increased so much?**

Renewable energy sources

If energy resources are to be managed in a sustainable manner, we must find ways to decrease the use of fossil fuels and balance this with an increase in the use of energy from renewable sources. In the search for renewable energy sources, it is important that the energy is captured in ways that are economically viable and at the same time avoid the worst effects of pollution. The costs involved may be in terms of the technology required to convert the energy or simply the transport costs incurred in bringing the raw energy source to the processing plant. To complement these developments, there is pressure to implement strategies for energy conservation and more efficient use of energy and thus minimise the escalating demand for energy.

Renewable energy sources (Figure 8.4) can be defined simply as 'energy flows which are replenished at the same rate as they are used'. A more comprehensive definition has been offered by the UK Renewable Energy Advisory Group as *'energy flows that occur naturally and repeatedly in the environment and can be harnessed for human benefit. The ultimate sources of most of this energy are the sun, gravity and the earth's rotation'.*

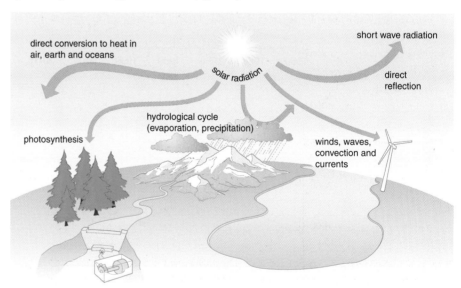

Figure 8.3 Solar radiation and how it is transformed into different forms of renewable energy

Theoretically, renewable energy sources are inexhaustible, though the practicality and technological expertise required to harness this energy on the scale required, as well as the economics of doing so, are areas that require furthur development and indeed are receiving active attention. Traditional uses of, say, wind power and to some extent fuel from biomass may have been adequate at the time, but there needs to be considerable expansion and development of renewable energy sources if they are to be appropriate for current and future global demands. An optimistic prediction suggests that, by the mid-21st century, about half the global energy requirements could be met from renewable energy sources.

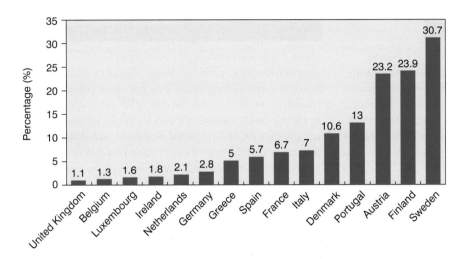

Figure 8.4 Renewable energy use in different European countries (as a percentage of the total energy use in the country)

Under the definition given above, we can now review the available renewable energy resources, some of which use solar energy directly whereas others harness solar energy indirectly. These indirect methods include the effect of solar energy on the weather, hence events occurring in the water cycle. The list below (and see Figure 8.3) summarises different renewable energy sources that can be exploited to provide energy.

- solar collectors for heating water (Figure 8.5a)
- photovoltaics, in which solar energy is converted directly into electricity (Figure 8.5b)
- wind energy (Figure 8.6)
- hydroelectric power (Figure 8.7a)
- wave energy
- tidal power (Figure 8.7b)
- geothermal energy
- conversion of **biomass** to **biofuels**.

Biomass is defined as the total mass of living material in an area (see Chapter 7). Growing plants utilise solar energy in the process of photosynthesis. Solar energy is thus incorporated into biomass, which then has the potential for use as a fuel by humans. In this section we explore some ways in which biomass is used as a renewable energy source and with this we can link the capture of energy from wastes, as much of this is itself derived from biomass. Crops grown specifically for use as an energy source are often described as **energy crops** and fuels derived from them may be described as **biofuels**.

ENERGY RESOURCES

EXTENSION MATERIAL

Some renewable energy sources

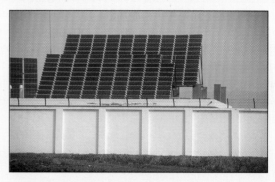

Figure 8.5a A simple (homemade) solar collector for heating domestic water (in Suffolk, eastern England). Water is pumped through a series of copper pipes clipped onto aluminium plates which are sprayed black. With good insulation behind the plates and a single pane of glass on top, the family living in this house find that, in summer, they rarely need to use any other energy source for their hot water supply.

Figure 8.5 (b) Photovoltaic (PV) solar panel unit in desert region (Xinjiang Province, northwest China). Cost of PV installations is high and relatively large areas of PV cells are required in relation to electricity yield. (An area of PV cells occupying the size of a football pitch could supply the electricity needs of a village in the UK with about 500 people.) PV technology is advancing, with promise of reduction in both cost and area required. Plans are being considered to cover vast areas of the Gobi desert with PV units, to make significant contribution to energy supplies of the future. Desert areas tend to have a high proportion of days with adequate solar energy!

(a)
(b)

Figure 8.6 Harnessing wind as a renewable energy source: (a) traditional windmill in Crete (Greece) being used to pump water for irrigation of spring crops; (b) Part of a modern wind farm near Urumqi in Xinjiang Province, northwest China. This is the largest wind farm in Asia. These wind turbines harness wind to generate electricity.

Figure 8.7 (a) Hydroelectric power (Elan Valley, Wales) – water in a river is retained by a dam, creating a lake behind it. Water then flows from the high level down a steep channel (known as a penstock) to the lower level where turbines in the hydroelectric station generate electricity (b) Barrage in the Rance estuary, France, designed to exploit tidal energy for generation of electricity. The barrage is positioned in an estuary. The incoming tide passes through open sluices but the

(a)
(b)

sluice gates are then closed at high tide. As the water falls with the ebb time, a 'head' of water develops across the barrage. This water is then allowed to flow through turbines to generate electricity. Relatively few schemes are in operation to harness tidal energy, but this system could make an appreciable contribution to UK energy requirements, particularly in locations with large daily tidal rise and fall.

Traditionally, wood is the main form in which biomass has been used as an energy source, though other materials such as animal dung are also used. The energy in the biomass is usually extracted by direct burning with a supply of air, in a stove to provide heat, or in a boiler to produce steam to drive turbines for generating electricity. In Britain, most of the commercial production of wood from the growth of trees in the forestry industry, is directed into timber for use in construction of buildings and carpentry in other ways. When a conifer plantation is felled, there is considerable waste in the form of small branches, needles and low-grade timber and this can be diverted for use as fuel.

Fast-growing biomass

Fast-growing energy crops that are currently used or are being developed through research programmes include:

- woody species, as short rotation coppice (SRC), e.g. willow, poplar
- non-woody species in the form of fast growing perennial grasses, e.g. *Miscanthus*, switch grass.

Short rotation coppicing (SRC)

The practice of coppicing has been used for several hundreds of years as a technique for management of woodland, particularly in Europe (Figure 8.8). (A description of traditional coppicing is given in Chapter 9.) Short-rotation coppicing (SRC) uses the same principle as traditional coppicing but with a reduced cutting cycle (Figure 8.9). In Britain, the species most commonly used are willow and poplar, because they are easy to propagate and fast-growing. Varieties are selected to give high yields. Cuttings are planted closely, then at the end of the first year they are cut off (coppiced) near the ground. Both willow and poplar then produce several side shoots. The crop of wood from these shoots is harvested after 2 to 5 years growth. The coppice is harvested by machines which cut and chop the material. Wood can probably be harvested from these stools for about 30 years, then the land would be returned to other agricultural crops. Short-rotation coppicing is an attractive option to some farmers as an energy crop.

Small plantations can be integrated within an existing farming system. It may be an appropriate way to exploit surplus land which, because of government controls, has been taken out of cultivation for food crops, such as cereals. Creation of coppiced areas with willow and poplar has benefits for wildlife, particularly in an arable farming area, because they provide an attractive habitat for birds and an increased diversity of plant species appears along the paths and within the coppice.

Fast-growing non-woody species

Miscanthus is a perennial bamboo-like grass, originating from Asia. This grass grows in clumps and, for example, *Miscanthus giganteus* can reach a height of 2 to 3 metres in one season. In European conditions it can produce high yields – 15 to 20 tonnes per hectare, though expectations are that this may go higher. The thick woody stems are cut down in the autumn, at the end of the growing season. The harvested material is baled and stored until required. In the following spring, fresh shoots grow from underground rhizomes, so a crop can be harvested annually. It probably takes 3 or 4 years for the grass to give its maximum crop, and it then can continue to yield for perhaps 15 to 20 years.

Figure 8.8 Traditional coppicing – the woody growth is generally cut ('coppiced') on a 10 to 15 year cycle. These ash stools may be hundreds of years old.

Figure 8.9 Short-rotation coppicing: a way of exploiting fast-growing woody species as a renewable energy source. This photograph shows young willows.

QUESTION

Summarise some of the benefits of growing short-rotation coppice.

Figure 8.10 Harvesting three-year-old Miscanthus

Figure 8.11 Harvesting short-rotation coppice willow

Figure 8.12 Using waste poultry litter to generate electricity – poultry litter in fuel hoppers at a power station in Suffolk

Other grasses with potential for use as fast-growing biomass include switch grass (*Panicum virgatum*) and reed canary grass (*Phalaris arundinacea*). It is likely that continued research will enable development of crops that are higher yielding and suitable for a variety of growing conditions.

Areas of land used in this way for growing energy crops are described as **energy plantations**. The harvested biomass 'crop' is made into wood chips or chopped so that is in a form that can be fed into a burner and converted, either to provide heat or for the generation of electricity. On a small scale, such as a single farm or cluster of houses as part of a community housing project, usually the most economical strategy is to use this energy to provide heating. Some farmers do use the wood chips produced on their own farms for running electricity generators; others sell the harvested crop to electricity generating companies as a fuel source. On a larger scale, however, the harvested biomass can be transported to a power station where it is used alongside other fuels for the generation of electricity. Such power stations may also be geared to utilise fuels such as chicken litter or wastes from forestry tree felling. Improvement in the design of the burners could lead to more efficient yield of energy from the crop in small-scale projects and thus reduce the cost (and energy used) in transporting the biomass to a power station.

There is considerable potential for exploitation of energy crops, particularly in rural areas, both in developed and developing nations. At the beginning of the 21st century, in parts of Europe (notably Germany and Denmark), established processing units for harvested biomass already make an appreciable contribution to the national energy use. Comparable projects lag behind in the UK but interest is growing and with appropriate government incentives (mainly in the form of grants), there is optimism that there will be a significant expansion of fast-growing biomass as a renewable energy source on a commercial scale.

QUESTIONS

Why do you think burning of stubble has been banned?
What is the main carbon compound in straw that gives off energy when burnt?
What benefits to the soil would there be from burning the straw on the fields?
In what other ways could surplus straw be used?

EXTENSION MATERIAL

Using straw as a source of energy

Straw left over after harvesting cereal crops is an example of biomass that has potential as a fuel source. In Britain, about 14 million tonnes of straw are produced annually. Some is used as bedding or for feeding livestock, but about half of the straw produced remained as an unwanted surplus. It had been the practice for surplus straw to be burnt along with the stubble in the fields at the time of harvest, leaving the field clear for planting the next crop. However, burning of stubble has been banned since 1992, so farmers have had to find other ways of disposing of the straw. Burners that can take straw bales have been developed and are used in some countries, notably Denmark and the USA. In the UK, now in 2003, a straw-fired power station is in operation near Ely, Cambridgeshire (eastern England). Compared with coal and oil, straw has very low levels of sulphur, but the main difficulties with utilising straw arise from the cost of collecting, storing and transporting a rather bulky material.

Gasohol from sugar

Fermentation of sugar by yeasts (*Saccharomyces* spp.), can convert the energy in biomass into ethanol which can be used as a fuel. An example is the fuel called **gasohol** which consists of 80 to 90 per cent of unleaded petroleum spirit with 10 to 20 per cent ethanol, and is used in motor vehicles. Most of the ethanol produced for gasohol uses sugar crops as its source material, though other plant species (such as maize and manioc), and waste materials (including wood and animal products), are also used.

Sugar is obtained from two different crops: sugar cane (*Saccharum officinum*) and sugar beet (*Beta vulgaris*). Sugar cane is grown in tropical and semi-tropical countries, whereas sugar beet is grown in more temperate regions. In the cane sugar industry, the cane is cut then processed. The initial milling yields two products: cane juice and bagasse (a residue of fibrous material). The juice can be further treated to the stage when sugar crystallises out, leaving a viscous sugary liquid known as molasses. Cane juice, bagasse and molasses can all be fermented to produce ethanol. Cane juice must be processed soon after harvest because it cannot be stored, whereas the molasses can be stored and fermented at a later date. Production of ethanol tends to be seasonal, linked to the time of harvest for the crop. Usually, small distilleries are set up at sites close to the crop. Bagasse is mainly cellulose, hemicellulose and lignin, and is often used to fuel the boilers used in the distillation process.

> The alcoholic fermentation of sugar by yeasts is represented in the following equation:
>
> $$C_6H_{12}O_6 \rightarrow 2C_2H_5OH + 2CO_2$$

EXTENSION MATERIAL

Gasohol and its potential

There is considerable potential for large-scale industrial production of ethanol for use as gasohol, but the decisions regarding its development are economic and political as well as environmental. Sugar cane and sugar beet are already grown in a wide range of climates and countries. Production of sugar crops could be expanded but at the risk of competing with other food crops. Ethanol has the advantage of being a relatively clean fuel, producing less pollution than the petrol it would replace if used as gasohol in motor vehicles. However, production of ethanol for gasohol is uneconomic compared with conventional petrol and would thus require both political and financial backing if it is to make a major contribution to the fuel used. Sugar cane growers may prefer to sell their crop to be refined as sugar rather than to be processed as ethanol. An extensive gasohol programme was initiated in Brazil during the 1980s. Yields of biomass from sugar cane are high in Brazil, and because of lack of available funds at that time there were difficulties in purchasing oil from overseas. However, a shift in oil prices or other political changes could alter the balance away from ethanol production and its development as a renewable fuel source.

Other biofuels

Certain vegetable oils, obtained by crushing seeds, can also be used as **biofuels**, without fermentation. Their energy content is similar to that of diesel, and higher than that of ethanol. Such oils can be blended, for example, with diesel fuel to contribute up to 30 per cent, though some need further processing to prevent clogging of the engines. Rape-seed oil (Figure 8.13) has been used in this way in Britain and the fuel produced is known as **rape methyl ester** (**RME**). Other crops that yield suitable oils include coconut oil (used in the Philippines), palm and castor oil (in Brazil) and sunflower oil (in South Africa).

*Figure 8.13 Other crops have potential as renewable energy sources: bright yellow oilseed rape (*Brassica napus*), grown for the harvest of oil from its seeds, can be used to produce a fuel known as biodiesel. The oils can also be used in industrial processes instead of oils derived from fossil fuels.*

Biogas from agricultural and domestic wastes

Biogas is a gaseous fuel which consists mainly of methane. Fermentation by bacteria can convert the energy in biomass to biogas. This process exploits the metabolic activities of different groups of bacteria which digest organic matter under anaerobic conditions.

A typical fermentation would produce biogas with a composition of about 65 per cent methane, 35 per cent carbon dioxide and traces of ammonia, hydrogen sulphide and water vapour. The methane in biogas burns with a clear flame to produce carbon dioxide and water without any hazardous air pollutants.

The process is used mainly with dung or slurries from animals so has the added benefit of turning waste material into a useful product. After digestion, the residue has value as a fertiliser. In Europe, the process is of particular value in the Netherlands and Denmark because there is a problem in finding enough land to spread waste pig slurry (liquid manure).

The waste materials contain mainly carbohydrate with some protein and lipid. The rather complex digestion process involves both aerobic and anaerobic bacteria and falls into three stages:

- **hydrolysis** – of the carbohydrate, protein and lipid, converting them to simple sugars, amino acids, and glycerol and fatty acids. This stage is carried out by aerobic bacteria.
- **acetogenesis** – as the available oxygen is used up, acetogenic bacteria convert the sugars and other substrates to acetic acid and other short-chain fatty acids and some carbon dioxide and hydrogen.
- **methanogenesis** – occurs only in anaerobic conditions due to the activity of methanogenic bacteria. These bacteria convert the acids to **methane** (CH_4). These bacteria are 'obligate anaerobes', which means they are active only in anaerobic conditions.

It is essential that conditions are anaerobic for the digestion to produce methane. For successful operation, temperatures are usually maintained between 30 and 40 °C. The methanogenic bacteria are sensitive to temperature changes and if fluctuations of more than 5 °C occur, the material goes sour due to build up of undigested volatile acids.

The digestion process is carried out in an enclosed tank, called a **digester**. The design of the digester may depend on locally available construction materials, but essential features are that it is strong enough to hold a large volume of the material to be digested and withstand the build-up of pressure inside. It must be gas-tight and allow the anaerobic conditions to be maintained. It should have an accessible inlet for loading the material, an outlet for the gas and a means of recovering the residue when digestion is completed. A simple but effective domed model, used widely in China, is shown in Figure 8.14. In some simple digesters, the vessel is sunk in the ground, thus helping to provide both support and insulation with respect to temperature. When digestion is completed, access to the digester pit is through the slurry reservoir. This allows the residue to be taken out and the reservoir cleaned. Often several of these small digesters are used together to ensure continuous supplies of gas. Larger sized industrial digesters are now becoming more widespread. Use of anaerobic digesters to produce biogas for energy has economic benefits and also reduces local water pollution and emission of greenhouse gases (see Chapter 9).

Biogas has been produced on a small scale in China for more than 50 years. Other countries such as Nepal, India and developing countries in Africa and South America also find biogas useful for small-scale production of fuel, particularly in rural areas, where it also provides a way of disposing of animal wastes and human excreta. In Britain, biogas digesters (Figure 8.15) are being used increasingly as a means of disposing of the large quantities of animal wastes derived from intensive farming methods. Ideally, the digester is located close to the source of biomass (slurry from cattle or pigs, or chicken litter) to avoid transport costs.

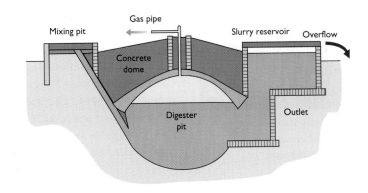

Figure 8.14 Fixed-dome biogas plant, of a type used, for example, in China and Nepal. As the fermentation progresses, gas given off collects in the dome and slurry is displaced into the reservoir. As the gas is drawn off and used, the slurry flows back from the reservoir into the digester pit.

EXTENSION MATERIAL

A closer look at the workings of a biogas plant

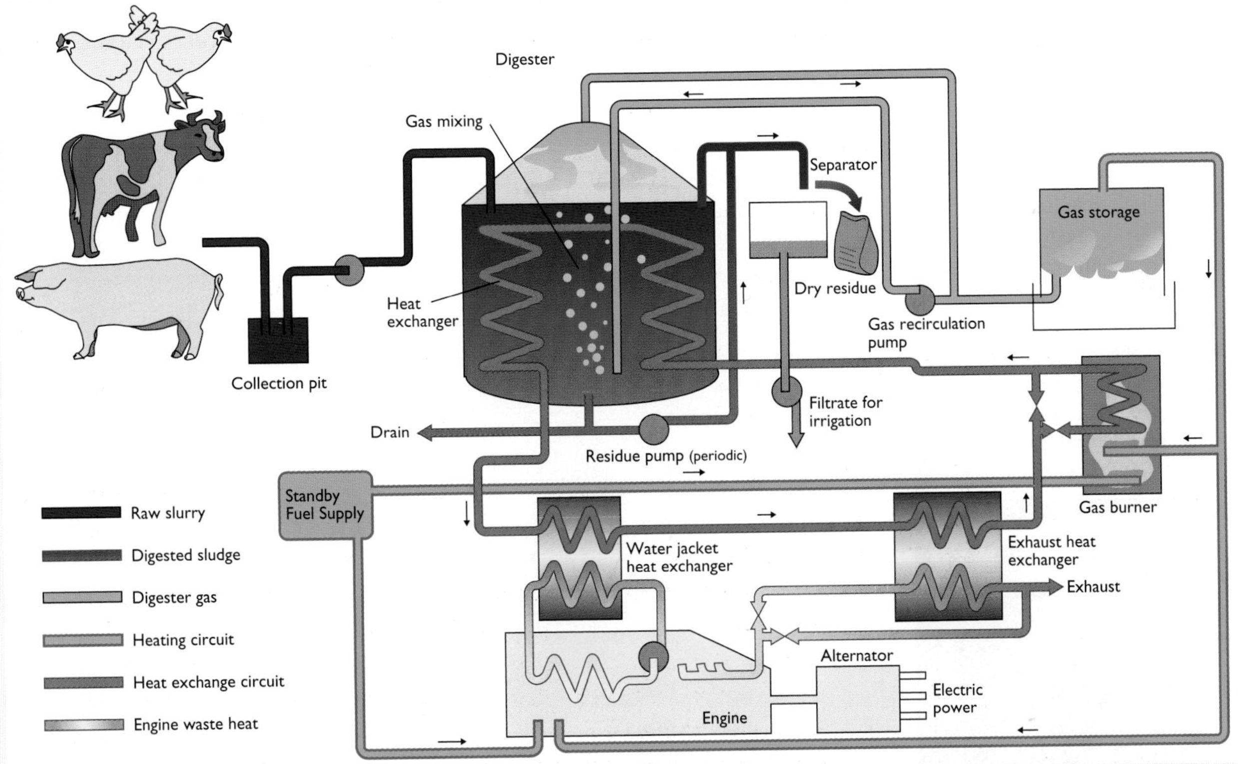

Figure 8.15 A biogas plant designed for use on farms in Britain. The biogas can be used to power an engine which generates electricity. In the digester, a suitable temperature is maintained by internal heating coils and heat exchangers, with biogas as the source of heat. Waste heat from the engine cooling system and exhaust can also be utilised, with a standby fuel supply for emergency. The contents of the digester are mixed by recirculation of the gas produced.

Figure 8.16 Mountains of domestic and industrial waste: most is dumped in landfill sites

QUESTIONS

List ways that biomass is used as a renewable energy source. As headings for your list, you could use:
1. agricultural wastes
2. energy from refuse
3. energy crops

In each case, indicate how far the source needs to be processed before the energy is transformed into a form that can be readily used.

QUESTIONS

In some areas in the UK, domestic households are requested to separate their rubbish into different bins (Figure 8.18). Materials that can be made into compost go into a separate bin from other waste.
What are the environmental benefits of separation? What can be recovered from the compostable materials and what about the rest?

Figure 8.18 Separation of domestic waste: the brown bin is for biodegradable materials which can be converted into compost

Rubbish or energy? – recovery of energy from landfill sites

Disposal of domestic and industrial waste has become a very large problem. In the UK, approximately 20 million tonnes of waste are produced each year, but the total figure is much higher (approaching 500 million tonnes) when all commercial, industrial, sewage and other wastes are included. These mountains of waste give rise to a number of environmental concerns. Dumping the waste occupies considerable areas of land and there is danger of leakage of toxic substances and other pollutants from the dumps. However, the waste material represents a source of energy which could be converted to fuel.

In Britain, nearly 90 per cent of domestic waste is disposed of in landfill sites (Figure 8.16). Biodegradable materials in the deposited waste (mainly paper, garden waste and foods) start to decay and soon use up available oxygen. Conditions thus become anaerobic, leading to the production of methane. The processes of hydrolysis, acetogenesis and methanogenesis are the same as described for production of biogas on pages 158 and 159. The gas produced is usually known as **landfill gas** when produced in the landfill site but, as for biogas, landfill gas consists mainly of methane with some carbon dioxide. In older landfill sites, the gas tended to seep out from the layers of compacted rubbish and there was the potential danger of the gas igniting or causing explosions. Since the late 1980s, an increasing number of landfill sites in the UK have been constructed to allow for recovery of the landfill gas. The energy in the landfill gas is then utilised by direct burning for heat, for use in engines or for electricity generation. A landfill site begins to produce gas about a year after landfill is completed and may be viable for up to 15 years (Figure 8.17).

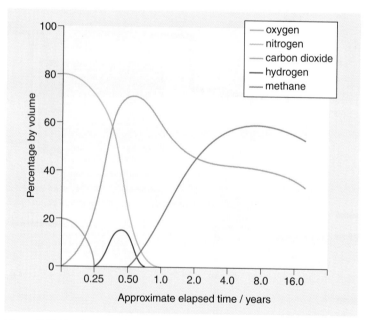

Figure 8.17 Production of gases in a landfill site and how the composition changes over time. Note the exponential time scale on the horizontal axis

EXTENSION MATERIAL

Aims for management of waste in the context of sustainability

As a means of reducing the energy loss and environmental damage through dumping of large amounts of waste, the following hierarchy outlines typical waste management strategies. Targets at national and local government level (e.g. in the UK) are in operation to help achieve these aims.

- reduction in proportion of waste going to landfill
- reuse of materials (such as glass bottles or vehicle tyres by re-treading).
  Note that there may be issues in terms of energy use and costs to put these schemes into practice. With any form of reuse, there must be an outlet for recycled materials and the processing must be economically viable and environmentally acceptable.
- recovery – in particular:
  - recycling of materials (such as glass bottles – but see note above regarding 'energy costs')
  - composting of organic materials in waste (as an extension of small-scale garden composting) plus use of sewage sludge as composting material
  - energy recovery – either by burning in an incinerator and recovering the energy by means of a boiler (to give hot water or steam to generate electricity) or by anaerobic digestion in a landfill site to produce the gaseous mixture known as landfill gas (see above)
- disposal – mainly in landfill sites (with or without energy recovery)

9 Human influences on the environment

Human activities, changes in land use and the impact on ecosystems

Early humans had relatively little impact on their surrounding environment. They gathered food plants, hunted animals, took small quantities of materials to build shelters and used wood as fuel to provide warmth and energy for cooking. As settled agriculture developed, their sphere of influence widened and humans began to change the environment in a more permanent way. Even so, as long as their activities remained on a small scale, disturbance of ecosystems was minimal and an equilibrium was more or less maintained.

Now, in the early 21st century, the scale of activities has escalated to a stage where the influence of humans on their environment has become overwhelming. There is considerable concern that some of these activities will lead to irreversible changes to our planet Earth as we know it today. Pressures on the environment arise from increased population and the demands associated with an increasingly complex and technological way of life.

As a result of human activity, over much of the globe natural ecosystems have been converted into agricultural land for production of food (Figure 9.1a). Agricultural land is likely to have low biodiversity because large areas of land are often dedicated to growing a single crop (known as a monoculture). This

(a)

(b)

Figure 9.1 Changes in land use lead to loss of natural ecosystems. The land shown in these photographs would have been covered in forest before the changes brought about by human activities: (a) use of land for agriculture – fields and hedges in England (left), terraced hillside in Nepal (right); (b) use of land for urbanisation – buildings, roads and car parks.

means that relatively few species are found on cultivated land. Further land is absorbed into living space in villages, towns and large sprawling cities (Figure 9.1b). In addition, loss of land occurs through development of industrialised areas. Land is also given over to the transport network and there is further disturbance of natural ecosystems due to exploitation of minerals through mining activities. Any change in land use is likely to alter the ecosystem that would naturally be present in the area. Here we look at **deforestation** and **desertification**, as two specific examples of how natural ecosystems have been altered or destroyed through the activities of humans.

Deforestation – its causes and effects

For thousands of years humans have invaded forests, cut down the trees and converted the area into agricultural land, to be used for growing crops or grazing animals. Often the debris has been burnt as part of the clearing process. In Britain, without human influence, much of the landscape would be dominated by woodland, whereas now only about 10 per cent of the land remains wooded (Figure 9.1a). A similar situation is evident worldwide. Agricultural expansion during the second half of the 19th century led to large-scale deforestation in temperate regions, but in the late 20th century there were particular concerns over the loss of tropical forest. Between 1960 and 1990, estimates suggest that about one fifth of all tropical forest was lost. The areas which suffered greatest losses were in Latin America, Asia and Africa. During the decade 1990 to 2000, estimates indicate that a further 2 to 4 per cent of forest cover has been lost. Another way of putting it is that an area approaching 12 million hectares (about the size of England) has been disappearing from tropical forests each year. Amazonia, mainly in Brazil, remains the largest area of rainforest. On a global scale, estimates suggest that at least 161 million hectares of forest were lost between 1990 and 2000 and, of this, 152 million hectares were in the tropics. During this same decade, only about 36 million hectares of regrowth of forest has been recorded and of this, 10 million hectares were in the tropics. As the rate of deforestation accelerates, recovery of *natural* forest may be impossible.

On a small scale, loss of forest to traditional systems of shifting cultivation (often cleared by slash and burn methods) may be short-lived. After a few seasons the people who cleared the forest move on to a new area. Provided the human population is of low density, the forest regenerates, and in a few years is likely to recover. If the forest is managed successfully for its productivity (usually for timber), it remains reasonably stable. Controlled reafforestation replenishes the stock of trees, thereby maintaining the ecosystem.

The large-scale damage to forest ecosystems that has been occurring over recent decades is a result of mechanisation, leading to massive clearance on an unprecedented scale (Figure 9.2). Pressures on forests, well illustrated by the Amazonian rainforest, arise from world-wide commercial demands as well as smaller but significant demands from indigenous people. The forests are exploited for their timber and fuelwood, mined for minerals (such as manganese, tin and iron), and used as a source for extraction of medicines (by large pharmaceutical companies as well as local people) and to provide food.

DEFINITIONS

Deforestation describes the removal of trees and clearance of forest land, which is then converted to other (non-forest) uses.

Reafforestation is the planting of trees on land that previously had carried forest (say within the last 50 years, or within living memory) but the forest on this land has been removed by natural or human activity.

Afforestation is the planting of trees on land that previously (or recently – see above) had not been used for forestry.

Both afforestation and reafforestation contribute to the overall **reforestation**, which refers to the deliberate establishment of areas of forest, but not including natural regeneration of forest.

Figure 9.2 Deforestation in Sichuan province, western China (right); natural forest in Sichuan province in western China (left), one of the few remaining areas still inhabited by the giant panda. The natural habitat is being eroded because of deforestation, encroaching agriculture and spread of industrialisation.

Roads have been built to regions in the forest that were previously inaccessible except to the indigenous people. Rivers have been utilised to generate electricity through construction of hydroelectric schemes. In Central America, areas of forest have been destroyed to provide grassland for 'ranching' of cattle and overpopulation has encouraged semi-permanent settlements of smallholders to encroach upon forest land.

Effects of deforestation

An immediate effect of deforestation is on the soil. The main effects can be summarised as:

- loss of nutrients
- deterioration of soil quality
- increased erosion.

Trees and associated vegetation tend to act as a sponge, retaining water and releasing it slowly to the soil and streams, and as water vapour to the air. With removal of the forest this stops abruptly; the soil surface loses its roughness and surface run-off of water increases. Large quantities of soil particles and nutrients are washed into streams and rivers. In tropical rainforests, the soil itself is relatively poor in nutrients, though considerable quantities are locked away in the biomass of the forest plants. These become available to future plant life through natural recycling. Removal of forest products, such as timber, takes large quantities of nutrients away from the area. Disturbance to the soil and removal of surface organic litter upsets the microbial population and the natural nutrient cycling mechanisms. The remaining soil is thus relatively poor in quality for any future agricultural use.

Sudden exposure of the soil means that the surface is heated up; it may become desiccated and is also subjected to the direct effects of rain and wind. The soil in the area is then more liable to suffer from **erosion**, which becomes particularly acute in sloping areas. Erosion can also affect more distant locations through deposition of sediment in rivers, which may lead to flooding (see below).

Hydrological effects

Deforestation can affect both water flow on the ground and water balance in the atmosphere. (Details of the hydrological (water) cycle are given in Chapter 7.) In previously forested areas, as well as actual erosion of soil, the flow of water into streams (and eventually rivers) is likely to be much greater. This is partly because removal of tree cover means that rain falls directly onto the soil and is not impeded or caught up in the branches and leaves of the trees (the tree canopy). It is also because removal of roots in the soil means that less water is taken up into the plant, and the physical barriers to water run-off have been removed. Increased water flow, together with deposition of sediment in rivers (from erosion) can lead to flooding, often at locations quite distant from the site of deforestation.

Removal of forest, particularly on a large scale, is likely to affect water balance in the atmosphere. On a simple level, the forest cover may encourage more rainfall (precipitation) in the area and certainly helps maintain high humidity in the atmosphere due to transpiration and evaporation (evapotranspiration) from the tree canopy. Evapotranspiration has a cooling effect. However, this cooling effect is lost when the area becomes bare, leading to a greater fluctuation in soil temperatures (see above). These effects may be considerable on a local scale but how far they really affect global climate is difficult to quantify.

> **DEFINITION**
>
> The term **evapotranspiration** refers to the loss of water vapour through the combined processes of evaporation and transpiration.

Global climate

The change from a forest ecosystem to agricultural crops or grassland for grazing may alter the utilisation of carbon dioxide in photosynthesis. If less carbon dioxide is used by the replacement crop, there will be a net increase in carbon dioxide in the atmosphere. Some of the changes may have long-term effects on global climate. The possible contribution to the greenhouse effect is discussed later in this chapter.

Biodiversity – species in the forest and their products

Probably the greatest resource of tropical forests lies in their biological diversity. Any attempt to estimate numbers of species is bound to be conservative, as many have never been recognised or described. Many of these unknown species are likely to be invertebrates, especially insects. A few examples from the Amazonian forest will give a hint as to their diversity.

> **DEFINITION**
>
> An area is described as having a high **biodiversity** if it contains a wide range of species, habitats and ecosystems.

QUESTION

Global change is not new and some of the concerns today are echoed in this passage written by Plato, 2300 years ago.

There are mountains in Attica which can now keep nothing but bees, but which were clothed not very long ago, with fine trees producing timber suitable for roofing the largest buildings ... . There were also many lofty trees, while the country produced boundless pastures for cattle. The annual supply of rainfall was not lost as it is at present, through being allowed to flow over the denuded surface to the sea, but was received by the country ... where she stored it ... and so was able to discharge the drainage of the heights into the hollows in the form of springs and rivers with an abundant volume and a wide territorial distribution ...

How far are these words, written by Plato about Greece, still true today? What effects does deforestation also have on species diversity?

EXTENSION MATERIAL

Diversity of animal life in Amazonian tropical forests

There are exotic birds, mammals and insects, often hunted for various trade outlets. Among the birds are toucans, parakeets and trogons; the monkeys include capuchins, howler and squirrel monkeys. There are anacondas, perhaps 10 m in length, boa constrictors and many other snakes, lizards, turtles, crocodiles, toads and frogs. The mammals include tapirs, anteaters, jaguars and capybara.

A range of products from tropical forests have direct value to humans as food sources. Many vegetables and fruits eaten around the world originate in tropical forests. These include rice, maize and potatoes, pineapples, avocados and bananas, coffee and cocoa, palm oil and Brazil nuts. Plants with medicinal value are used by the indigenous people, and some have been used in modern medicine. Examples are quinine for malaria, and curare as a muscle relaxant in surgical operations. Curare is obtained from the bark of trees and for centuries was used by South American Indians as an arrow poison.

Biodiversity – loss of species and of genetic resources

It is difficult to predict the long-term effects of the loss in biological diversity as forests are destroyed. Inevitably species are lost. Some species may adapt to life elsewhere, but others depend critically on factors in the forest ecosystem, say a food plant or the shelter provided. The giant panda exists in only a few locations in China, being largely dependent on bamboo for food, but its survival is seriously threatened by the destruction of forests in these crucial areas (Figure 9.2). Endangered species from other forests include apes, lemurs, elephants, leopards, tigers, some parrots and crocodiles.

Another potential threat to biodiversity occurs when small areas of forest, which have not been cut (felled), become yet smaller as trees continue to be cut from the edge of the remaining patches. These isolated fragments of forest may then become too small to support the full range of species of the former forest. One way to minimise the effect is to establish 'corridors' between the different fragments and thereby encourage movement at least of animal species between these remaining areas of forest. This strategy is being adopted in Sichuan, western China, which is one of the few areas that still supports populations of the giant panda.

The genetic reserves in both animal and plant species in the rich forest ecosystem are potentially highly valuable resources. The genetic variation in commercial breeds of domesticated animals and in crop plants used in agriculture has become dangerously narrow, with the risk that whole populations could be destroyed by disease or by pests. Wild populations may, therefore, provide sources of fresh genetic material, with the potential of introducing favourable characteristics, such as flavour and improved disease resistance, or allowing the crop plants to be grown in different climatic regions.

The influence of humans is not a new problem. In prehistoric Britain there were large mammals: lions, leopards and hyenas, similar to species currently living in

Africa, and more recently there were bears and wild swine. The demise of these large mammals was probably due to the combination of hunting and loss of the protective cover of native woodland as it was cut down. The last known aurochsen (ancestors of modern domesticated cattle) were observed in forests in Poland in 1627. Reduction of woodland to the point where surviving animals could be hunted and caught almost certainly contributed to their extinction.

The impact of deforestation can be devastating on indigenous forest people. They lose the wood as a source of fuel for cooking; they lose the trees as a source of food and income, but above all, their cultural and spiritual homeland is destroyed. Money and offers of social housing are unlikely to restore this or adequately recompense the people affected.

An example of unsustainable exploitation of a forest area is seen in parts of Borneo where, over the last 30 years, the tropical forest has been progressively cut down (about 12 per cent disappeared from one area during the 1980s). This deforestation is being carried out in favour of logging operations and for establishment of rubber and oil plantations. The indigenous Dayak people have lost their source of food and medicine, of clothing and shelter. The effects of deforestation are seen in the degradation of the soil, silted streams, floods and droughts and a reported reduction in biodiversity in this area.

Forests and management for sustainability

Increasing awareness of both the short-term and the long-term effects of deforestation has led to the development of strategies designed to halt or reverse the level of global deforestation. Such strategies aim to encourage sustainable use of existing forest resources and also to promote deliberate planting of trees, in areas previously forested (reafforestation) or those which had not recently been forest land (afforestation). These measures are often government backed, and many large areas of forest now have management plans. These plans aim to ensure that there is control over the cycles for cutting (felling) the trees, followed either by natural regeneration or by replanting schemes. The plans also take into account the extent of exploitation for other purposes. They must be monitored rigorously at a local level (often in remote areas of a country) to ensure they are being implemented successfully.

EXTENSION MATERIAL

Recovery from deforestation – using China as an example

In China, in the early 1980s, it was predicted that, if the rate of forest depletion continued at the then current rate, China would be virtually out of wood by the year 2000. In the early 1990s, in the southwest mountainous regions of China, heavy logging traffic was evident. However, by 1999, roadside checks ensured that logging loads have valid permits, though some illegal felling certainly continues to occur. At the same time, there was widespread evidence of tree planting schemes, both in rural and urban locations. Indeed, many Chinese cities show the benefit of urban tree planting schemes. Some mountain areas, which had been ravaged in earlier decades to produce fuel to boost industrial output, are now supporting young forest cover. Such forests provide fuel for immediate local use as well as timber in the longer term. Some afforestation schemes are linked to stabilising of agricultural land by reducing soil erosion. One requirement of the 'Obligatory Tree Planting Programme' (adopted in 1981) is for all Chinese citizens over the age of 11 years to plant three to five trees each year or to make a contribution to afforestation work. With these and other plans, China aimed to create 66 million hectares of new forest by the year 2000, an increase of about 20 per cent in its forested areas.

Coppice woodland – an example of sustainable management

The term **sustainability** is referred to in Chapter 8. Maintenance of forests for sustainability requires management. The management should allow products of the forest to be harvested and the forest should continue to be productive and also to support a diversity of both flora and fauna.

Figure 9.3 Coppiced woodland as an example of sustainable management: Bradfield Woods (in Suffolk, eastern England) in late summer. The foreground shows 2 to 3 year growth from coppiced stools, interspersed with full height standards deeper into the woodland

Maintenance of forests for sustainability requires management if the forest is to allow harvest of its products and continue to be productive and also to support a diversity of both flora and fauna. This can be illustrated by management systems used in coppice woodlands, such as those found in Britain. Coppicing is a method of woodland management that has been practised for many hundreds of years. The wooded areas are usually laid out in blocks (known as fells) and the coppicing or cutting of the woody species takes place in these blocks. The blocks are often separated by wide grassy tracks, with perhaps a shallow ditch each side. At intervals, certain tree species (such as alder, ash, birch, chestnut, hazel, hornbeam, maple, sallow, small leaved lime – or others, depending on the location) are cut down close to ground level. New side shoots then grow out from the stumps (known also as stools) and are allowed to grow for, say, 10 to 15 years, then are harvested as poles (Figure 9.3). Different blocks are coppiced in successive years, giving a crop of wood every year. Among the coppiced stools, a few trees (usually oak, but also species such as ash, silver birch, elm or willows) are allowed to grow to their full size as mature trees. These are known as standards. When felled, these trees supply timber, used in building houses. Trees regenerate naturally from seedlings. The continual regeneration of trees from the coppiced stools maintains a supply of wood in a sustainable way.

Coppicing allows a diversity of plant and animal species to be supported within the woodland community. Comparison of the ground flora in different coppice areas reflects stages in a succession, from open ground through to areas with dense shading from the tree cover. Open rides offer sites for other flowers and along the edges of the rides there can be shrubs and taller vegetation, providing attractive habitats for insects, including butterflies. Ditches, if present, introduce yet another habitat for a different range of species. The different stages of the coppicing cycle create a range of habitats, hence a diversity of species. Bradfield Woods (see Extension material) gives an example of management in a coppice woodland and illustrates how suitable management can contribute to sustainability in terms of continued harvesting of the crop or other products. At the same time, this example shows how the management helps to maintain biodiversity in the woodland area – indeed, many of the species listed in the description of Bradfield Woods are rare or absent in the surrounding farmland and would disappear if the woodland were to be destroyed.

QUESTIONS

The description of Bradfield Woods gives an example of how management of a woodland by coppicing allows harvest of its products.

- What evidence is there that this can be considered a *sustainable* management system? What evidence is there that *species diversity* is maintained or enhanced in Bradfield Woods, compared with the surrounding area?
- Think of another example, in your local area if possible, that illustrates a sustainable system. Look at how it is managed, what the products are and any evidence of good species diversity.
- Try to work out a concise definition for the term *sustainability*, that could be applied to forest management and to agricultural systems.
- How far is sustainability realistic for agricultural systems on a global scale?

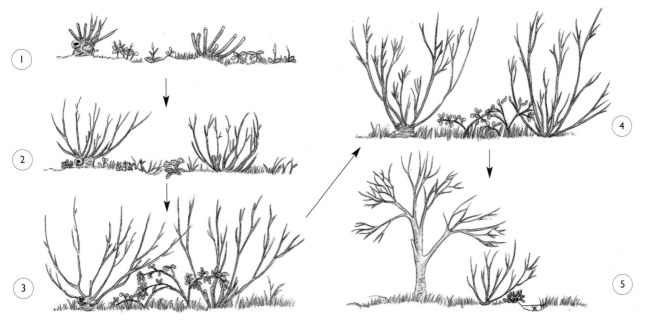

Figure 9.4 Management of woodland by coppicing helps to maintain biological diversity of species – some stages in the coppicing cycle in Bradfield Woods (Suffolk):

Stage 1: first year after coppicing. Ground species include dog's mercury and water avens. Herb Paris and early purple orchid may be seen, both indicators of ancient woodland.

Stage 2: second and third years after coppicing. Richest ground flora seen at this stage, in spring. Ground species include sweet violet, bugle, wood anemone, primrose, dog's mercury, patches of oxlip (now rare), bluebell and wood spurge.

Stage 3: fourth and fifth years after coppicing. More grasses, brambles and other species invade.

Stage 4: eight and nine years after coppicing. Ground flora more sparse, and has lost its colour and diversity (though the cycle starts again when the area is freshly coppiced).

Stage 5: climax woodland stage.

EXTENSION MATERIAL

Biodiversity in Bradfield woods – an example of coppice woodland

Bradfield woods is an ancient coppice woodland in Suffolk (UK). (Further details about Bradfield woods are given in *Genetics, Evolution and Biodiversity*, Chapter 5.) In Bradfield woods, some of the ash stools reach a diameter of up to 6 m, with an estimated age of at least 1000 years (see Figure 8.8 – photograph also shows Bradfield Woods). They have probably yielded crops of wood continuously over this period. In the history of Bradfield Woods, the harvested wood has been used for the manufacture of various products, including wooden rakes, handles for scythes, thatching pegs, hazel for daub and wattle (used in local timber-frame buildings), fencing materials and hurdles as well as firewood.

The series of coppiced areas, representing different cutting intervals within the 10 to 15 year cycle, show successive stages of regrowth of the shoots from the stools (Figure 9.4). In Bradfield Woods, at least 42 native trees and shrubs have been described, about two-thirds of the total in Britain. Over 350 species of flowering plants are known from these woods, including a number of rarities. However, if left to grow without any management, the diversity of species within the wood would gradually diminish. Hazel is an example of a woody species that does not compete well in dense, mature woodland. Similarly, in Bradfield Woods, as well as the abundant plant life, there are plenty of birds and the abundant birdsong includes that of willow warblers, blackcaps and nightingales. Other animal life includes adders, grass snakes, frogs and toads and a range of typical small woodland animals.

Desertification – its causes and effects

Natural deserts occur in both hot and cold regions. They are characterised by low and intermittent rainfall, usually less than 250 mm of precipitation per year. Sometimes total drought persists from one year to the next. Because of the low water availability, desert areas support only limited vegetative cover.

The term **desertification** has been defined as 'land degradation in arid, semi-arid and dry sub-humid areas arising mainly from adverse human impact'. This definition recognises that *human activities* are the main cause of desertification occurring at the present time and associates its progress with increased population and consequent pressure on the natural resources of the area. Climatic variations (particularly reduced rainfall) may increase the rate of deterioration caused by human activity and make the situation even worse. Concern arises about the loss of productivity from land which could be used for agriculture, leading to worsening poverty or even famine among the people the area has traditionally supported.

Pressures on the land come mainly from grazing flocks (often sheep, cattle or camels) and from gathering fuelwood. Other pressures include deforestation (see page 165) and deliberate burning of vegetation by local people. To this we can add the effects of wheel tracks from motor vehicles and trampling (by people and livestock), particularly in the area close to settlements. Some arid areas are only just able to support the growing of crops, probably helped by irrigation, but the situation is usually already fragile. A dry season can be disastrous and result in crop failure. Any additional pressure, from increased human population or worsening of the drought conditions, means that grazing of animals and fuel gathering spreads over a yet wider area. Demands for grazing by the herds may be greater than can be provided by that particular environment or area of land. We can say that the demands exceed the 'carrying capacity' of the land. (More details about carrying capacity are given in *Genetics, Evolution and Biodiversity*, Chapter 4). The vegetation cannot support the grazing herds, nor can it recover unless the pressure is removed (Figure 9.5).

Figure 9.5 Desert areas in Afghanistan: (a) sparse vegetation; (b) grazing pressure; (c) fuel gathering; (d) erosion in area denuded of vegetation

Effects of desertification

Lack of vegetation leads to a downward spiral in terms of deterioration of the land. The bare soils, exposed to direct sunlight, suffer further desiccation and, combined with exposure to winds, are more likely to suffer from **erosion**. Soil particles may be deposited elsewhere, perhaps over other marginal crops or grazing land, causing further loss of potential agricultural land. The wind-blown soil and drifting sand also damage transport systems by blocking roads and railways, and often threaten irrigation channels and rivers. This disrupts irrigation systems or may cause flooding, perhaps in locations some distance away from the area affected by desertification.

High rates of evaporation in the absence of plant cover are likely to alter the water–salt balance. Salts drawn up from the ground water are left behind in the surface layer of the soil. This leads to **salinisation** which contributes to the deterioration of the soil, which then becomes unsuitable for plant growth (see Figure 9.6(a)). Salinisation may also result from poorly designed irrigation systems, where the existing drainage is unable to handle an increased water supply. As the groundwater level rises, dissolved salts are brought to the surface, but the area effectively becomes waterlogged so the salts remain in the water. With high rates of evaporation, the salts then accumulate at the soil surface. The situation may be exacerbated if chemicals have been used on the land, either as fertilisers or for pest or weed control.

Loss of vegetation inevitably reduces biodiversity. The overall effects are detrimental to people living in the area with respect to loss of land for grazing and for the production of crops, leading to greater poverty in the affected areas. Lack of vegetation reduces the amount of moisture in the atmosphere, which may in turn affect the climate pattern in the area, particularly rainfall.

Two examples described here emphasise how human pressures on the land have brought about desertification in areas where an equilibrium had previously existed. Disaster struck the Sahel region of Sudan in the early 1970s. This semi-arid zone had had a higher than average rainfall for about 20 years. This encouraged greater cultivation in the region, including cash-crops, such as peanuts. The people whose livelihood depended on their grazing herds were pushed further north into the fringe of the Sahara desert, crowded into land with already poor carrying capacity. When the drought returned, thousands of people and millions of animals died because they had no food, inadequate water and nowhere to go.

In the second example, in northern China, extensive areas of semi-arid grassland have suffered from desertification, seen as deterioration of the land with more sand blowouts and shifting sand dunes. Increased human population in the area is linked to political movements of people. Agriculture that is settled in one place has been replacing traditional practices, which depended on seasonal movement of grazing animals. Since 1949, in an area of Inner Mongolia, the number of livestock has increased while the land available for grazing has decreased (see Table 9.1). Breeds of cattle, such as the European Friesian and Simmental, were introduced. These have potentially higher productivity, but there has not been the necessary adjustment in herd size to take account of their higher food intake.

Table 9.1 *Grazing pressures and desertification: Xilingole League, China (1949 to 1980)*

Year	Number of livestock	Grazing area / km²	Density / number km⁻²
1949	1 740 000	193 000	9.01
1958	4 750 000	193 000	24.63
1964	8 210 000	179 000	46.00
1970	5 780 000	123 000	47.04
1980	5 270 000	142 000	37.02

Collection of fuelwood to support the increased population has worsened the situation.

Recovery from desertification

Plans to reclaim or rehabilitate land that has undergone desertification, or to prevent further degradation of threatened land, require commitment by governments and understanding at the local level of the people in the affected communities. At government level, there needs to be a strategy that sets out what should be done, sufficient organisation to implement the necessary steps and funding to enable the activities to be carried out. An important part of any strategy must be to involve local people and ensure they are educated and trained so that they understand what needs to be done and why.

Generally a first important step is to re-establish vegetation cover by planting trees to act as a windbreak. Within the protected area, the likelihood of further degradation is reduced and there can be a gradual recovery by growing crops and also by establishing grass or other fodder crops for grazing. Strict control of grazing on a rotational basis and distribution of the livestock according to the carrying capacity of the land allows good use to be made of limited resources in these areas. Once the initial steps have been taken on a small scale, reforestation (by both afforestation and reafforestation) can be undertaken on a larger scale. In time, this relieves the pressure on fuel gathering and lifts the area from being one in which the people are struggling to survive.

A variety of strategies is used when farming in arid areas to try to overcome the difficulties associated with minimal rainfall. Slopes are often terraced to prevent run-off and trees can be planted to create shade and wind shelter for crops. Leguminous tree species can be planted to increase soil fertility. Stones left on the ground trap moisture, reduce evaporation and create shade and shelter for seedlings. Stones can also be placed around the base of trees to help retain moisture. In sub-Sahara, research showed that survival of wild shrub seedlings depends on the proximity to stones – in the first summer 90 per cent survive if found within 0.5 cm of the edge of a stone, but rates are much lower if further away. In Australia, farmers are using new techniques to restore exposed, hard-baked areas suffering from degradation. Bulldozers are being used to cut a series of shallow grooves and depressions in the ground. These artificial hollows act as traps for moisture and for wind-blown seeds. The small hollows then shelter the seedlings, enhancing rooting and increasing chance of survival.

Other strategies for prevention of desertification and spread of desert are being adopted in Iran, which has over 5 million hectares of sand dunes. In certain areas the dunes are sprayed with oil which stabilises the sand surface. The coating of oil prevents seedlings from being blown away and retains moisture. After spraying, drought-resistant shrubs are planted in a series of narrow furrows, cut across the direction of the wind. One area of more than 60 km² was treated in this way, and, in 6 years was transformed from bare sand to a forest of trees more than 3 m high.

A project in northern Ethiopia illustrates how careful management can provide an effective mechanism for environmental rehabilitation in desertified areas. In this project, certain areas of bare land are selected then enclosed to protect them from grazing. This allows the original vegetation to regenerate. To improve conservation of soil and water, the land is terraced, reinforced by stone along the banks and paths, and trees and grasses are planted. When the vegetation has become established sufficiently, there is a gradual return to gathering of the grasses and the wood from within the enclosed area. This is known as the 'cut and carry' system. The grass may be used as fodder for animals and as roofing material for the houses. To ensure fairness and control, at all stages the decision making is done by members of the community. Agreements are drawn up as to which areas of land are to be enclosed, when 'crops' may be harvested and which households are eligible to benefit in any particular year. In one such area of about 35 hectares (at Alasa, northwest of Tigrai), these principles have been applied, enabling recovery from precarious subsistence farming on marginal land to a stabilised system, yielding crops, fodder and fuel. There is also a noticeable increase in wildlife (including bees for production of honey) and clear evidence of a significant reduction of problems with erosion and accumulation of silt.

(d)

(a)

(b)

(c)

Figure 9.6 (a) Patchy and poor quality crop growth in an area suffering from salinisation, at the edge of a desert oasis (Xinjiang Province, northwest China). (b) Trees at the edge of oasis cultivation in a desert area in Xinjiang Province, northwest China. Trees are extremely important in maintaining these fragile areas of cultivation – they act as a windbreak, helping to protect the crops from encroaching wind-blown sand and gravel. They also provide valuable shade and wood (for buildings and fuel). (c) Stabilisation of sand dunes in Mauritania, as a first stage in reclaiming land for cultivation. Windbreak fences are made from branches of Balanites aegyptiaca. Balanites *is also used for making furniture, oil is extracted from the fruit, and drugs are made from the roots and bark. (d) Trees being planted in Mali, to act as windbreaks and help reclaim desert land.*

Figure 9.7 Atmospheric pollution from industrial activities: discharge from a sugar beet factory (in eastern England)

Pollution and its effects

Pollution is a familiar word in our everyday vocabulary. We are repeatedly reminded of how a range of human (anthropogenic) activities produces harmful substances which contaminate the air, the water and the land which together make up the environment for living organisms. Pollutants are derived mainly from the waste materials produced by industrial activities (including agriculture) and from motor vehicles, with some coming from domestic sources (Figure 9.7). The consequences of pollution are observed in the effects on living organisms and, on a different scale, in the possible long-term effects on climate.

Atmospheric pollution

The term **atmospheric pollution** refers to change in the constitution of the atmosphere brought about by human activities, causing harm to humans or to other living organisms in the environment. Now, in the early 21st century, atmospheric pollution affects all nations of the world. Pollution has increased over the last 150 years and this increase is linked to the increasing human population and the rapid growth of urban and industrial societies. However, pollutants travel in air currents and spread across international boundaries so the effects are not confined to urban or industrial areas.

Damage caused by atmospheric pollution affects people, their crops, buildings and wildlife as well as the global climate. There is considerable concern that some of the effects are irreversible because the atmosphere has limited ability for recovery. With increasing awareness of the effects of pollution, some changes have been made in the relevant human activities, but a realistic reduction can only be brought about by efforts involving individual, local, national and international controls. The following sections look at the causes and consequences of two aspects of atmospheric pollution: increasing acid rain and the enhanced greenhouse effect.

Acid rain

Rain, snow and other forms of precipitation are naturally mildly acidic, with a pH of about 5.6. (Remember that pH 7 represents neutral pH and that change by one unit on the pH scale represents a 10-fold change in acidity.) The natural acidity of rain is due to carbon dioxide in the air, which reacts with water to form carbonic acid. We use the term **acid rain** to describe precipitation of pH 5 or lower.

EXTENSION MATERIAL

The history (and future) of acid rain

Here we look at different methods that have been used to trace changes in acid rain over time and, in some cases, have helped to identify sources of the polluting gases.

Evidence for past changes in acid precipitation can be found by analysing samples at different depths in glaciers and ice sheets. For thousands of years, up to the start of the Industrial Revolution, the pH was near to 6, or even higher, whereas by the middle of the 20th century, pH values of 4 to 4.5 have been commonly recorded in North America and Europe. Another technique uses evidence from deposits of diatoms in the sediment of lakes. Diatoms are microscopic organisms and their cell walls are impregnated with materials containing silica. Because this is hard, the remains of different diatoms are preserved as fossils and the depths at which they

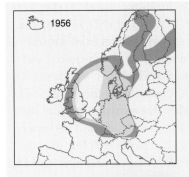

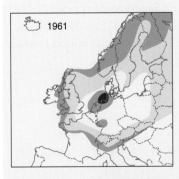

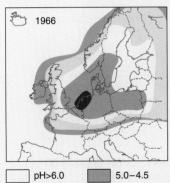

pH>6.0	5.0–4.5
6.0–5.5	4.5–4.0
5.5–5.0	<4.0

Figure 9.9 Increases in the acidity of precipitation in Europe between 1956 and 1966

occur provide a useful record of past history. Diatoms have preference for certain pH ranges, so the profile of diatoms deposited in sediments in certain lakes can give an indication of the history of the pH in the lake (Figure 9.8).

Over a shorter time-scale, detailed monitoring of the acidity of rain in Europe over the 10-year period from 1956 to 1966 showed that the rain had become more acidic and that the area affected had expanded (Figure 9.9).

More recent data (Table 9.2) looks at records of sulphur dioxide emissions from burning fossil fuels in Asia, Europe and the United States from 1980, with predictions to 2010. These data clearly indicate a downward trend in certain industrialised nations, but emphasise the concern over increased emission of polluting gases from developing nations in Asia and consequent threat from acid rain in these areas.

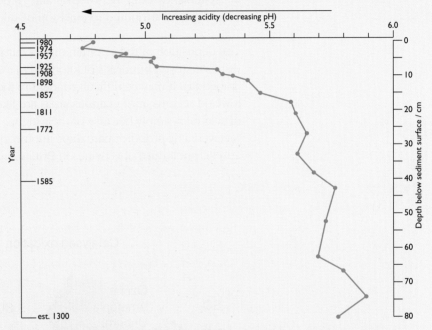

Figure 9.8 Fossil diatoms can give an indication of the history of the pH in a lake. These data show the analysis of diatoms from Round Loch, a lake in southwest Scotland. Despite its isolation, the lake shows sharp increases in acidity from about the time of the Industrial Revolution in the mid-19th century.

Table 9.2 *Sulphur dioxide emissions from burning of fossil fuel in Asia, Europe and the United States, 1980 to 2010. Sulphur dioxide emission is linked directly with deposition of acid rain and these data show the increasing threat of acid rain now posed by developing nations in Asia*

Region	Million tons of sulphur dioxide in different years				
	1980	1990	1995	2000*	2010*
Europe	59	42	31	26	18
United States	24	20	16	15	14
Asia	15	34	40	53	79

* predicted values

This increased acidity (see Extension Material) is attributed mainly to the production of sulphur dioxide (SO_2) and oxides of nitrogen (NO_x) during the burning of fossil fuels. A small proportion of these gases arises from natural events, including volcanic eruptions, microbial decay of organic material and lightning (NO_x). Events leading to the formation of acid rain are summarised in Fig 9.10. Sulphur dioxide and oxides of nitrogen released into the atmosphere can simply fall to the ground. This is known as **dry deposition**. In addition, a complex series of reactions may take place, involving oxidation and becoming dissolved in water suspended in the atmosphere. The gases (sulphur dioxide and oxides of nitrogen) are thus converted to acids – mainly sulphuric acid and nitric acid. These acids are deposited on the ground in rain, fog, hail, sleet and snow and this is known as **wet deposition**. Both dry and wet deposition are included under the term **acid rain** and both have damaging effects on the ecosystems. Globally, between 60 and 70 per cent of the acid deposition that occurs can be attributed to emission of sulphur dioxide.

The deposition of the acid rain may occur in areas that are considerable distances (even hundreds of kilometres) away from the source of pollutant. Sometimes it may be difficult to identify the exact source. Compared with lowland regions, mountainous areas are likely to receive a relatively high dose of acid rain, simply because of their higher rainfall. Scandinavia appears to have received a disproportionate deposition of acid rain. Some of this probably arises from industrial activities in Britain, but a substantial contribution is made

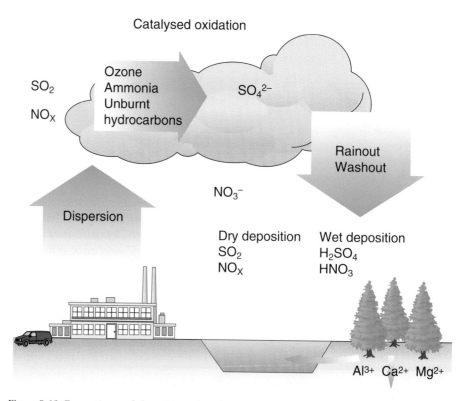

Figure 9.10 Formation and deposition of acid rain

by industries in continental Europe, including the eastern European countries. During the 1950s, the policy in Britain was for the introduction of tall chimneys for power stations. This was an attempt to reduce ground level pollution by lifting the emissions to a higher level so that they were dispersed away from the source. While at least partially successful on a local scale, this practice has undoubtedly contributed to acid rain deposition at more distant localities.

The effects of acid rain

The effects of acid rain on animal and plant life in both aquatic and terrestrial ecosystems, on buildings, and to some extent on human health are now well recognised.

EXTENSION MATERIAL

Acid rain and its effects on lakes in Norway

In the late 1970s, a study made of 1679 lakes in Norway, showed a strong relationship between the pH of the lakes and the abundance of fish in them (Figure 9.11). Most of the lakes with a pH below 4.5 had no fish whereas virtually all those with a pH higher than 6.0 had good populations of fish. Similarly, there is evidence of loss of fish, because of the increased acidity, from rivers in Canada and the United States of America, and in lakes in Scotland and Wales.

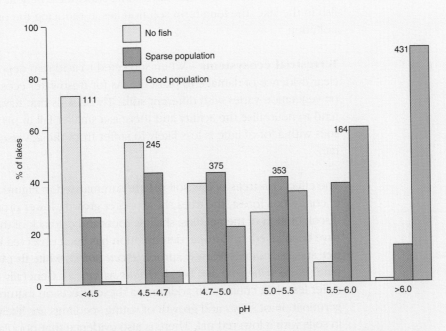

Figure 9.11 The relationship between pH in the water and abundance of fish. These data were obtained in the late 1970s from 1679 lakes in Norway and show how low pH (high acidity) appears to have a detrimental effect on fish. The numbers above each group of bars indicate the number of lakes observed within the particular pH range.

Freshwater ecosystems – There is plenty of evidence to link acid rain deposition with a decline in fish populations and other freshwater organisms. Lowering of the pH affects fish in various ways and some examples are given below.

- Acidity leads to reduced calcium concentration. It also encourages mobilisation of certain heavy metals that are otherwise locked away in clay particles and the bedrock beneath the water. They become soluble and are released into the water where they may accumulate to toxic levels. These include aluminium, cadmium and mercury. As an example, aluminium appears to make fish produce an excess of sticky mucus on their gills. This leads to reduced intake in salt through the gills, which interferes with the osmoregulating process of the fish. Because of the mucus, the gills become clogged and gas exchange becomes difficult.
- Below pH 4.5, trout do not produce the enzyme that breaks down the outer coating of eggs, so the larvae get trapped inside. This prevents successful reproduction.
- Loss of phytoplankton (See Background Material on page 185) in acidified water has made such lakes noticeably more transparent. This is because there is less microscopic material in suspension. The effects spread through the food chains – some species become more abundant but others are lost. In some lakes where the pH is below 6, the moss *Sphagnum* shows vigorous growth, covering the bottom of the lake and pushing out other plant life.

Attempts have been made to reverse the effects of acidity by adding lime to affected lakes. While this may have some short-term benefit to the stock of fish in the lake, the long-term solution lies in reducing the origin of the pollution.

Terrestrial ecosystems – When subjected to acid rain deposition, there is clear evidence of damage to plant life. As for freshwater ecosystems, the impact on vegetation varies with different soils. Those soils that have a buffering effect tend to neutralise the acidity and there is a smaller fall in pH. Vegetation on soils with a lot of lime is less likely to suffer from the adverse effects of acid rain.

The effects on trees of low soil pH are summarised in Figure 9.12. In large areas of coniferous forest, the trees show poorer growth, lower productivity, discolouration of the needles, shallow roots and die-back of the crown. Many have been killed. In Europe, deterioration has been observed in some important tree species, notably Norway spruce (*Picea abies*), white fir (*Abies* sp.), Scots pine (*Pinus sylvestris*) and beech (*Fagus sylvatica*). Generally deciduous trees suffer less than conifers. In addition to these effects on mature trees, germination of seeds and growth of young seedlings are likely to be inhibited in soils with a lowered pH. There is also evidence that populations of animals have declined in terrestrial ecosystems affected by acid rain. Reduction or absence of other soil organisms, including microorganisms, is certain to interfere with the natural processes of decomposition and nutrient cycling in the soil.

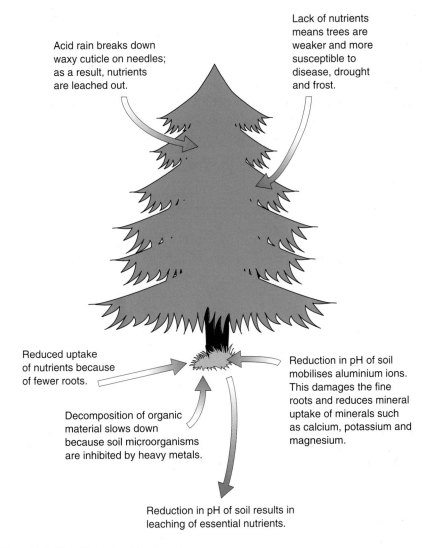

Acid rain breaks down waxy cuticle on needles; as a result, nutrients are leached out.

Lack of nutrients means trees are weaker and more susceptible to disease, drought and frost.

Reduced uptake of nutrients because of fewer roots.

Reduction in pH of soil mobilises aluminium ions. This damages the fine roots and reduces mineral uptake of minerals such as calcium, potassium and magnesium.

Decomposition of organic material slows down because soil microorganisms are inhibited by heavy metals.

Reduction in pH of soil results in leaching of essential nutrients.

Figure 9.12 The effects of acid rain on trees

Human health – Acid rain may also affect human health. A low pH in the soil releases ions of certain heavy metals, such as cadmium, lead and mercury, and these may contaminate drinking water supplies. Pregnant women are at risk from water originating from affected areas, because the fetus is particularly sensitive to mercury poisoning. Acidity may cause leaching of copper and lead from water systems. In Sweden, high levels of copper and lead have been detected in drinking water. Copper from this source may have caused outbreaks of diarrhoea in young children and account for green colouring to baths and even hair! Release of aluminium may be harmful because of the possible link between aluminium and Alzheimer's disease (pre-senile dementia) in humans.

Other more general health problems that have been reported from areas affected by acid pollution include increases in respiratory illness and, particularly in children, increased frequency of colds, coughs and allergies.

QUESTIONS

Suggest ways that acid rain can be reduced – on a local as well as on an international scale. How can *control* of acid rain be monitored? [*You may find some help at the end of this chapter.*]

The greenhouse effect

The greenhouse effect (Figure 9.13) is a natural phenomenon in our global atmosphere and plays an important part in maintaining life on Earth (as we know it). Without a greenhouse effect, global temperatures would be some 30 °C lower. In Europe, for example, summers would be more like winter and the winter temperatures would approach those in the Arctic and Antarctic regions. There would be much less liquid water on the surface of the Earth, tropical rainforest would not exist and the crops that could be grown around the world would be severely limited.

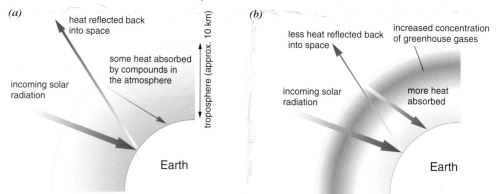

Figure 9.13 Summary of events that contribute to the greenhouse effect. Note that the troposphere is the lowest layer of the atmosphere and extends to a height between 9 and 16 km above the surface of the Earth. (a) water vapour and other gases absorb heat, which warms the troposphere. Heat is radiated back and warms the Earth; (b) greenhouse gases, such as carbon dioxide, methane and chlorofluorocarbons (CFCs) absorb more heat, so less is reflected back into space. This increases the temperature of the troposphere.

BACKGROUND MATERIAL

Understanding the greenhouse effect

To understand the greenhouse effect, we need to trace the pathway of the energy received from the sun, through the atmosphere of gases around the Earth to the surface of the Earth. Radiation from the sun is emitted in a range of wavelengths. Those that are important to biological systems range from short-wave ultraviolet (less than 400 nm), through visible (400 to 700 nm), to infrared (more than 700 nm). Gases in the atmosphere are relatively transparent to this incoming radiation, although about 30 per cent is reflected back into space. Part of the ultraviolet is absorbed by the reactions between oxygen and ozone so the ultraviolet radiation is effectively filtered out in the stratosphere. Some of the remaining energy that passes through the atmosphere warms the surface of the Earth. Energy is then radiated back away from the Earth's surface as longer wavelength infrared (4000 to 100 000 nm). Part of this escapes through the atmosphere into space, but some energy is absorbed by gases in the troposphere because the gases are less transparent to this longer wavelength infrared.

Events that contribute to the greenhouse effect are summarised in Figure 9.13 and in the Background material. Energy absorbed by the gases in the troposphere results in the troposphere warming up. This warm layer then re-radiates the heat energy it has gained. Some is radiated back to the Earth, again providing warmth. The effect of these gases in the atmosphere is to keep the surface of the Earth warmer than it would be without the gases. This way of trapping the heat is known as the **greenhouse effect**.

We can see a parallel (but not identical) situation in a greenhouse (glasshouse). The glass is transparent to incoming radiation, so both the ground and the air inside warm up. The panes of glass help retain the warmth, partly because they help prevent the warm air escaping by convection. The ground and air are thus kept warmer than they would be without the glass covering. The gases in the troposphere are equivalent to the panes of glass, though they retain the heat in a different way.

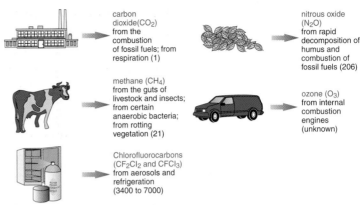

Figure 9.14 Sources of greenhouse gases. Numbers in brackets show the relative effectiveness as a greenhouse gas, per molecule of gas, compared with that of carbon dioxide. Water (vapour) also acts as a greenhouse gas and is the most significant greenhouse gas in the atmosphere. (No comparable figures are available to indicate the relative effectiveness of water).

Several different gases absorb the long infrared radiation and these are known as **greenhouse gases**. The main greenhouse gases are listed in Figure 9.14, together with their estimated contribution. With the exception of CFCs, all these greenhouse gases occur naturally, but all of them are also produced as a result of human activities. Increases in the levels of the greenhouse gases have become apparent over the last 150 years, alongside development of industrialisation. The rate of increase has accelerated in recent decades, as we move into the 21st century. The concern is that these anthropogenic (human) sources of greenhouse gases are leading to an exaggerated or enhanced greenhouse effect and that the acceleration will continue unless positive steps are taken to limit the accumulation of greenhouse gases.

The **carbon dioxide** level in the atmosphere is maintained primarily by the balance between respiration and photosynthesis. Levels of carbon dioxide at different times in the past have been estimated from carbon dioxide trapped as bubbles in ice in Antarctica and Greenland. These show a continuing and steady rise of carbon dioxide from a pre-industrial level of about 280 ppm (parts per million) to about 315 ppm in 1958 and 353 ppm by 1990. These figures show how the increase has accelerated in the three decades from 1960 to 1990 (Figure 9.15a). The annual fluctuations reflect seasonal changes in the rate of photosynthesis, but the trend is clearly upwards.

Increases in carbon dioxide are possibly due to the burning of fossil fuels (coal, oil and natural gas). In the early 1990s, this was estimated to be in excess of 6 billion tonnes per year. The carbon in fossil fuels was fixed by photosynthesis millions of years ago when the vegetation was growing, so burning the fuels now releases carbon dioxide which had effectively been removed from circulation. Deforestation on a large scale may also upset the contemporary balance between respiration and photosynthesis (see pages 163 and 167, earlier in this chapter). When forest trees are cut down and removed, the land is often converted to agricultural land which is then used for grazing or growing other crops. The amount of carbon dioxide used in photosynthesis

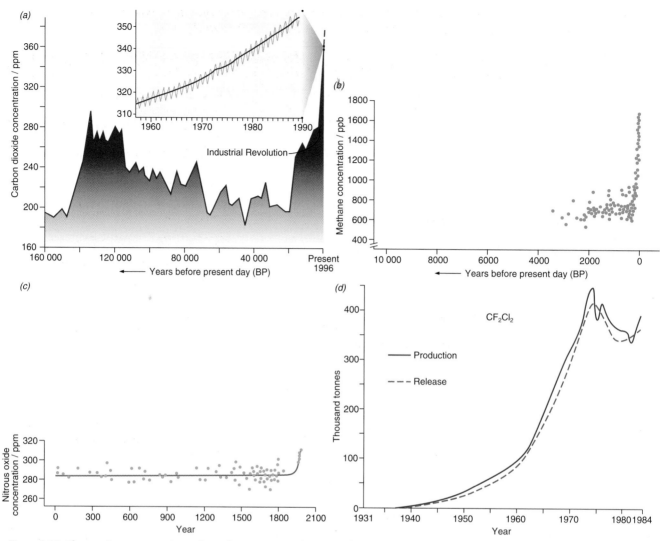

Figure 9.15 Changes in concentrations of greenhouse gases in the atmosphere (note the different time scales): (a) carbon dioxide over 160 000 years, (inset) carbon dioxide detail from 1956 to 1990; (b) methane over 10 000 years; (c) nitrous oxide over 2000 years; (d) CF₂Cl₂ (one of the CFC gases) over 50 years (from 1931 to 1984).

is likely to be very much less than when trees were growing, resulting in a net increase of carbon dioxide in the atmosphere. Clearing the land during deforestation involves burning of residues in the forest and disturbance of the soil may release further carbon dioxide. The main events of the carbon cycle are summarised in Chapter 7, Figure 7.16.

Further details relating to chlorofluorocarbons, methane, nitrous oxide and water vapour, as greenhouse gases, are given in Table 9.3

Table 9.3 *Some greenhouse gases (other than carbon dioxide). [The gas **ozone** also makes a contribution as a greenhouse gas, but is not included in this table.] (**see** also Figures 9.14 and 9.15)*

Gas	Sources	Other comments
Chlorofluorocarbons (CFCs) – e.g. CF_2Cl_2 *see Figure 9.15 (d)*	• aerosols • refrigeration	▫ produced only as a result of human activities ▫ much more efficient than carbon dioxide at absorbing infrared radiation ▫ particular concerns as they persist in atmosphere for 60 years or more *(see page 192 and Figure 9.22)*
Methane (CH_4) (also known as 'marsh gas') *see biogas and 'landfill gas' on pages 158 to 160* *see Figure 9.15 (b)*	• naturally by bacteria in wet, marshy, anaerobic conditions (e.g. rice paddy fields) • guts of ruminant animals (e.g. cattle) • landfill sites	▫ increasing demand for food means more rice paddy fields are cultivated and more cattle are reared ▫ rise in temperature would release more methane, now trapped in frozen tundra areas
Nitrous oxide (N_2O) *see Figure 9.15 (c)*	• naturally in soil by denitrifying bacteria • nitrogen fertilisers applied to soil • burning of fossil fuels and other biomass • decomposition of human and animal wastes	▫ enormous increase in use of fertilisers is significant in relation to relatively recent increase in nitrous oxide levels
Water vapour (H_2O)	• evaporation from the seas of the Earth • burning of fuels and other industrial processess	▫ rise in temperature would cause yet more water to evaporate from the sea ▫ most significant greenhouse gas

It is well established that levels of greenhouse gases have increased over the last 150 years. It is also recognised that human activities produce greenhouse gases and that these activities have increased during the same period. This suggests that, as a result of these increases in greenhouse gases, there may be an enhanced greenhouse effect, leading to a rise in temperature on the surface of the Earth (Figure 9.16). This is described as **global warming**. Climatic patterns since the mid-19th century do show a rise in global temperature as a general trend. We cannot, however, say categorically that the temperature rises have been *caused* by human activities, even though we accept that there are strong links between the two.

Events that determine climatic patterns are highly complex. Reconstruction of past climates shows that there have been considerable fluctuations in global temperatures at least over the past 20 000 years, dating back to the last glacial period. The recent rises in temperatures may be part of a general fluctuation, such as has occurred before. Extrapolation of present trends suggests, however, that during the early part of the 21st century, global temperatures are likely to rise to a level higher than at any time in recent history.

There is considerable concern about the implications of human activities and their possible effects on global climate but it is difficult to predict precisely what will happen in the immediate future. People may respond by reducing

> **QUESTION**
>
> What measures can be taken to reduce the contribution of chlorofluorocarbons (CFCs) to global warming?

those anthropogenic activities which cause emission of greenhouse gases. This could be supported by policies agreed on an international scale, though this cannot have an immediate effect because there has already been a build-up of certain persistent gases in the atmosphere.

If we assume that production of greenhouse gases continues at the present rate, a best estimate suggests that, by the year 2030, carbon dioxide concentrations will reach double the pre-industrial levels. Computer models predict that this doubling would lead to a global warming of between 1.5 and 4.5 °C or more. This may not seem a very large increase in temperature, but the effects could be far reaching. One certain result would be a rise in sea levels. This would be due partly to thermal expansion of the sea water in the oceans and partly to melting of glaciers and the ice sheets of Greenland and in Antarctica.

Climatic changes would also lead to changes in patterns of rainfall and temperature. There would probably be shifts in the distribution of both natural ecosystems and agricultural crops compared with their distribution today. Yields of crops might benefit from the higher temperatures and from higher carbon dioxide levels. The zones in which cultivation is successful may extend beyond their present limits. As an example, maize could probably be grown 200 km further north in Europe. However, weeds and crop pests would also benefit, and vectors of disease might alter their range, which in turn would

QUESTIONS

Why is it wrong to say that the greenhouse effect is *caused* by human activities? How far do you think that the increased human population has contributed to the enhanced greenhouse effect? Try to link specific human activities with the greenhouse gases they produce.

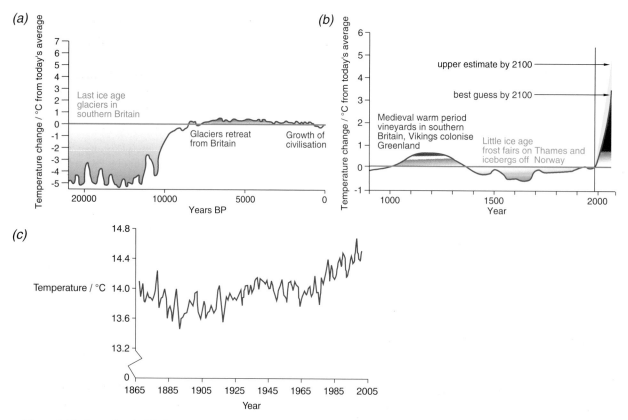

Figure 9.16 Generalised global temperatures: (a) estimated global temperatures during and just before the history of human civilisation, shown as a comparison with today's average temperatures; (b) detail of the last 1000 years, with prediction of global temperatures into the next century; (c) global average temperature at the Earth's surface, from 1867 to 2001.

affect crop losses. Perhaps with the warmer climate we would use less fuel to provide heating, but this could be offset by more air conditioning to help cope with the hot summers. We cannot predict the answers with certainty.

Water pollution

Water on the Earth continually circulates through the processes of evaporation, transpiration, condensation and precipitation. The main events of this **water cycle** are summarised in Figure 7.12 (see page 146). Human activities can interfere with or pollute the water at any stage of the cycle (Figure 9.17). Pollutants may alter the physical conditions in the water, thus disturbing the balance of organisms living in aquatic habitats, or they may do direct harm to living organisms which would normally be present. There is increasing concern over contamination of human drinking water, which is extracted from groundwater and also from rivers. Some pollutants, such as nitrates, have reached unacceptably high levels in these water sources in certain areas. Pesticide residues in water may have harmful effects and polluted water may also encourage spread of disease.

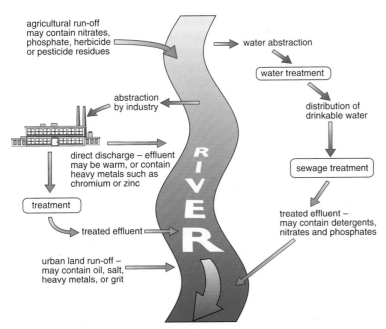

Figure 9.17 Events affecting quality of water in a river

Effects of raw sewage on water quality

Sewage from domestic sources consists mainly of human faeces and kitchen waste. The latter includes detergents and residues from food preparation. Sewage is also likely to carry some industrial waste, which may incorporate stronger liquids used in cleaning or in a variety of industrial processes. Other components may be present, including toxic substances. Examples here are chromium from leather tanneries or copper and zinc from metal-plating industries.

BACKGROUND MATERIAL

Aquatic organisms that may be affected by pollution

Plant life in water ranges from microscopic organisms to larger plants. The microscopic organisms are known as phytoplankton and include bacteria, blue-green bacteria, diatoms and other algae. Through the photosynthetic activity of its organisms, the phytoplankton is very important in the primary productivity of aquatic ecosystems (see Chapter 7). Population numbers fluctuate dramatically in response to seasonal changes or availability of nutrients. Large visible algae in the water include blanket weed (*Cladophora* spp.) and large plants include Canadian pondweed (*Elodea canadensis*) and water lilies (*Nymphaea alba*).

Animal life in water ranges from microscopic zooplankton, through invertebrates, fish amphibians, birds and mammals. The level of available oxygen is an important factor in determining the distribution of animal species. Certain invertebrate species can be used as indicators of pollution levels because some tolerate very polluted water with poor oxygenation whereas others are restricted to clean well-oxygenated water (see Chapter 3).

Some information on adaptations shown by organisms in aquatic habitats is given in Chapter 3 and this is helpful in understanding changes that occur in populations of organisms as a result of pollution.

In Britain and many other countries, much of the sewage produced is treated in sewage treatment plants and this involves both physical and biological processes. The **effluent** (liquid) and **sludge** (solid) produced after treatment is completed can usually be used or dumped without harm to the environment. However, sometimes raw sewage is released into rivers or discharged into the sea without treatment, giving rise to pollution. Occasionally, accidental leakage of sewage occurs, or its treatment may be inadequate. Pollution may also result from leakage on farms of liquid waste from animal manure (slurry) or from silage (stored and fermenting grass).

Suppose some raw or inadequately treated sewage or slurry is discharged into a river. There will be instant changes in the physical factors of the aquatic environment and an impact on the organisms in the river. Table 9.4 shows the main components of sewage and summarises some of the effects of such pollution in a river. By looking at the graphs in Figure 9.18, we can follow this in more detail through various stages downstream.

Table 9.4 *The main components of sewage and their effects in a river*

Component of sewage	Features	Consequences on environment
suspended solids	• size ranges from large and visible to colloidal and dispersed • usually organic, so are degradable and can be decomposed by microorganisms	□ reduce penetration of light □ high demand for oxygen during breakdown of organic material by microorganisms
nitrogenous compounds	• originate mainly from proteins and urea • often present in the form of ammonium compounds (NH_4^+) • oxidised in stages by nitrifying bacteria (see *Nitrogen* cycle) to nitrates:	□ NH_4^+ ions toxic to fish □ excess nitrate leads to **eutrophication** □ eutrophication leads to **algal blooms** □ respiration of algae leads to an increased **biochemical oxygen demand (BOD)**, resulting in depletion of oxygen □ death of masses of algae results in an increase in BOD (while the algae are broken down by aerobic bacteria) □ some toxins are produced during growth of algal bloom (from certain blue-green algae) □ high nitrate concentration is damaging to human health if the water is used as a source for drinking
phosphates	• present in faeces and modern detergents	□ excess phosphate leads to eutrophication (described above for nitrates)
toxins	• heavy metals such as Cu, Pb, Zn, may accumulate • persistent pesticides originate from agricultural run-off rather than sewage	□ toxic effects on organisms in the water, or for humans if the water is used as a source for drinking water
microorganisms	• may include viruses, bacteria, protozoa, and fungi (some may be pathogenic)	□ health risk for humans, particularly if the water is used for drinking without adequate treatment
detergents	• 'hard' detergents (used in the 1950s) create foam and are unsightly on the surface • 'soft' detergents (used since the 1960s) are biodegradable, but rich in phosphates	□ foams on the surface interfere with aeration of the water □ high levels of phosphate may lead to eutrophication

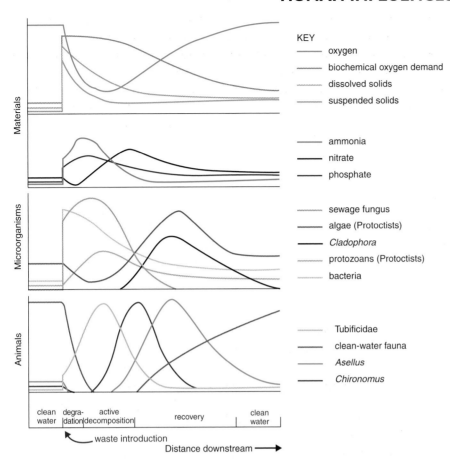

Figure 9.18 Typical changes in water quality and in plant and animal populations in a river after introduction of raw sewage (see text for discussion of events). The arrow shows the point at which sewage is introduced to the river. Above this point the water is considered to be 'clean'. Moving downstream from the arrow, we can identify different zones of microbial activity, linked to fluctuations in levels of oxygen, solids in the water, dissolved salts, microorganisms and other aquatic fauna. First there is a zone of degradation and active decomposition of material that was introduced into the water. This is followed by a zone of recovery before reaching clean water again. (See text for detailed discussion of events.)

In the initial stage, just below the point at which sewage enters the river, decomposition of the organic material occurs immediately. The organic material may include proteins, carbohydrates and fats as well as more complex particles of organic matter. These particles are sometimes suspended rather than dissolved in the water. This decomposition is mainly due to the activities of aerobic bacteria. These bacteria may have been present in the river water or associated with the sewage material. The demand for oxygen is high, resulting in a rapid decrease in the concentration of oxygen in the water. This 'oxygen sag' occurs when the rate of utilisation of oxygen in respiration by the microorganisms exceeds the rate at which oxygen is replenished. In extreme cases, the oxygen level may fall so low that conditions become anaerobic. In very polluted water, visible slime-like growths known as 'sewage fungus' are likely to appear. The term sewage fungus is misleading as the slime consists of filamentous bacteria, protozoa and algae (members of the Protoctists) as well as fungi. Sewage fungus is tolerant of high levels of ammonia and anaerobic conditions.

BACKGROUND MATERIAL

Oxygen and biochemical oxygen demand (BOD)

Oxygen dissolves in water and its concentration is a critical factor in determining the types of living organisms (plants, animals and microorganisms) that are present in the aquatic environment (see Chapter 3). The solubility of oxygen decreases with increasing temperature (see Figure 3.7). Organisms use oxygen in respiration and the oxygen level fluctuates on a daily basis, with an increase during daylight linked to its production by photosynthetic activity in plants. Seasonal changes in temperature and light intensity also influence oxygen availability because of their effects on photosynthetic activity.

The demand for oxygen varies considerably, depending on the number of organisms present and on their activity. The amount of oxygen being used by living organisms in the water is often expressed as the **biochemical oxygen demand (BOD)**. The BOD of a sample of water is measured under standard conditions at 20 °C over a period of 5 days. Presence of polluting organic material in the water leads to increased activity from microorganisms involved in the decomposition of this material and this results in a high demand for oxygen. Measurement of BOD can, therefore, provide an indication of the level of pollution.

As the organic substances in the sewage are used, the numbers of aerobic bacteria decrease, but protozoans that feed on other organisms show an increase. The intensity of light at different depths within the water is a critical factor in determining the distribution of photosynthetic organisms. Light intensity decreases with depth and is also reduced by material suspended in the water. As these suspended solids begin to settle out or decompose, the water becomes clearer. This means that light can penetrate to greater depths. This then allows an increase in the number of algae. These algae and aquatic plants carry out photosynthesis, thus helping restore the oxygen level in the water. Inorganic ions are released from the decomposing material. Ammonia, though initially at high levels, is soon converted to nitrate by nitrification. The nitrate, phosphate and other ions released from the material are likely to be utilised by the increasing populations of algae.

The lowest graph in Figure 9.18 shows typical fluctuations of invertebrate populations in the different zones of the river downstream from the sewage discharge. The *Tubifex* worms and *Chironomus* (midge larvae) are tolerant of low oxygen levels, whereas *Asellus* (water louse), snails, leeches and fish require progressively cleaner water (see Chapter 3). These population numbers reflect changes in the clean quality of the water and demonstrate how the water can effectively go through stages of self-purification as a result of the activities of microorganisms in the water. In highly polluted waters, there is a danger that recovery from the initial drop in oxygen concentration occurs only very slowly. The distance over which such purification takes place depends on many factors, including the temperature of the water, the severity of the initial pollution and the existing microbial population in the water. To some extent, the effect of the pollutant becomes less as it is diluted as it passes downstream.

Effects of fertilisers on water quality – eutrophication and algal blooms

Use of chemical fertilisers on agricultural land has increased markedly with changing agricultural practices aimed at increasing yields of crops (Figure 9.19). The main inorganic ions applied in fertilisers are nitrate (NO_3^-), phosphate (PO_4^{3-}) and potassium (K^+) (Figure 9.20). These ions dissolve in soil water and are leached out by rain and water percolating through the soil. The water drains into lakes or rivers giving them an artificially enriched nutrient status. In some cases the body of water may initially have had a relatively low level of nutrient, whereas in other situations the nutrient status may naturally have been higher. High nutrient levels are likely to support high productivity in

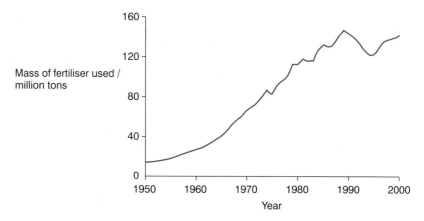

Figure 9.19 Global use of fertiliser, from 1950 to 2000. Note the marked rise over the first decades of this period, but signs of a reduction over the last decade.

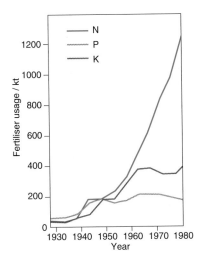

Figure 9.20 Increases in use of fertilisers containing nitrogen (N), phosphorus (P) and potassium (K) in the UK, between 1928 and 1980. Increased use of fertilisers is linked with increases in the concentration of nitrate in rivers.

Figure 9.21 Eutrophication in a lake in Kunming, southwest China. Excessive growth of water hyacinth fills channels leading into the lake.

BACKGROUND MATERIAL

Nutrient levels in the water and eutrophication

Statements about 'nutrient status' generally refer to levels of inorganic nitrogen and phosphorus, although other mineral elements are also required by aquatic organisms. Some water is relatively poor in nutrients. When water is rich in nutrients it is described as **eutrophic**. (The 'eu' in eutrophic means 'well' or 'good' so the word implies 'plenty of nutrients'.)

terms of biomass within the water environment. We use the term **eutrophication** to indicate the artificial nutrient enrichment of an aquatic system, regardless of its initial status (Figure 9.21).

Excess or enrichment of nutrients in the water encourages rapid growth of algae and blue-green bacteria, resulting in a population explosion known as **algal bloom**. Such growth may also occur in natural situations stimulated, say, by a seasonal increase in temperature. The phase of rapid growth may be followed by sudden and massive mortality, perhaps because the supply of one of the nutrients has become limiting. Dead algae then rise to the surface of the water as an unsightly green scum.

The most serious consequences of algal blooms arise from the depletion of oxygen. The growing algae carry out respiration and their demand for oxygen is particularly high at night. During the day, the net output of oxygen from photosynthesis is likely to be greater than that used in respiration. When the algae die, the mass of organic material is broken down by microorganisms (mainly aerobic bacteria), making further demands on the supply of oxygen. If the level of oxygen falls too low, there may be sudden death of masses of fish, a situation known as **fish kill**. Waters that are severely eutrophic may lack oxygen and under anaerobic conditions unpleasant odours become apparent. These are due to production of hydrogen sulphide. Under these conditions methane (marsh gas) may also be produced. Certain blue-green bacteria produce toxins that may also cause the death of fish. The bacterium *Clostridium botulinum* flourishes in anaerobic conditions. The toxins it produces cause paralysis known as botulism, which can affect birds and mammals (including humans swimming in the water). Overall, there is usually a drop in species diversity as well as reduction in numbers. Some species, however, prefer or can tolerate the lack of oxygen.

BACKGROUND MATERIAL

Movement of water and oxygenation

Movement of water is likely to increase its oxygenation, particularly when it is fast. Aeration is increased by turbulence, so shallow rivers flowing fast over a rocky base acquire more oxygen than stagnant pools. In a given stretch of a river, there is a continual exchange of water, so that events at some distance upstream can affect the river much lower down.

QUESTION

Sometimes, as a result of a discharge of raw sewage, a sudden mass mortality of fish occurs, known as fish kill. High levels of ammonia are toxic to fish and there may be poisonous substances such as heavy metals in the water. What is likely to be the main reason for the sudden death of fish?

Excessive growth of algae as a result of eutrophication restricts penetration of light into the water. Other water plants suffer from a reduced rate of photosynthesis and this further diminishes the supply of oxygen. So we can see how the events following eutrophication can lead to a rapid deterioration in the quality of water and changes in the populations of plants, animals and microorganisms. Recovery is generally associated with improvement in oxygenation as well as dilution of the original source of nutrient enrichment.

Even after quite severe incidents of water pollution, recovery and rehabilitation of the habitat often occurs. Usually the most important stage is for oxygenation to be restored and then the plant and animal species need time to re-establish and build up their numbers, though the species composition may have changed.

Pollution, people and politics

In an ideal world, once a link has been established between the effects of pollution and its origins, people would modify their activities in ways that would reduce or minimise the harmful effects of pollution. However, as we move into the 21st century, it is clear that we live in a society that is dominated by the use of fossil fuels, with very widespread use of motor vehicles. At the same time we live in a 'throw away' economy. Substantial changes are needed in personal lifestyle as well as in the industrial sector if we are to make a realistic impact, on a global scale, on controlling and reducing both the current levels of pollution and the demands on natural resources.

The prime targets for reduction are emissions of carbon and of sulphur, use of pesticides and artificial fertilisers, and the use of 'virgin' raw materials. The challenge to do this has been taken up on an international as well as a local scale and it is through legislation that the intended strategies are implemented. In the interests of political survival, no government can be too extreme in the measures it lays down, because individuals in society are rarely willing to accept a lowering of their standard of living. A balance needs to be found between those measures that are practical to enforce and those that are realistic in relation to the economic needs of the society. This balance must be acceptable on a local as well as a global scale. Some examples from past and recent legislation in the UK and the European Union (EU) are used here to illustrate legislation that has been drawn up with the aim of controlling levels of pollution in both air and water.

EXTENSION MATERIAL

A view of the past

Even in 13th-century Britain, the adverse effects of smoke from burning coal were recognised and there were penalties, some severe, for using 'sea coal' in open furnaces. The mid-19th century really saw the beginning of modern pollution control measures. This was in response to pressure from the public over effects of smoke linked to bronchitis and damage to the countryside from the pollution near industrial plants. The Smoke Nuisance Abatement (Metropolis) Act of 1853 and the Alkali Act of 1863 were early attempts to curb the damaging effects of industrial smoke. A century later, the Clean Air Acts of 1956 and 1968 introduced smoke control areas which gave a measure of control over domestic and industrial use of fuels and encouraged greater use of smokeless fuels.

There has been a noticeable decrease in smoke emission linked to the Clean Air Acts (see Extension Material), but in the 1970s there was concern over the continued high levels of sulphur dioxide and of other pollutant gases, such as carbon monoxide, oxides of nitrogen and ozone. During the 1970s, there were European Community (EC) Directives, which attempted to lay down standards relating to air quality. These, for example, set out maximum permitted concentrations of smoke and of sulphur dioxide, and one aim of these directives was to protect health. Within the EC, there was a requirement for a network of air pollution monitoring stations to ensure compliance with the standards set. Other measures included controls over motor vehicle emissions and phasing out of leaded petrol.

As another example of legislation, concern about the effect of CFCs on the ozone layer led to a treaty (1987) aimed at reducing their use and a noticeable reduction followed. Since 1996, production of CFCs has been banned in industrialised nations and there is a commitment for production in developing countries to be phased out by the year 2010. CFCs contribute to the enhanced greenhouse effect (see pages 183 to 185) as well as causing damage to the ozone layer.

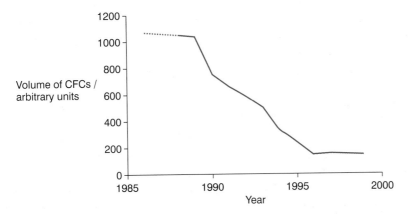

Figure 9.22 World production of CFCs, from the late 1980s through to 2000. Note the remarkable drop in response to the 1987 treaty

European legislation and the control of air and water quality

In the late 1990s, the EU defined and established objectives for air quality control, to 'avoid, prevent or reduce harmful effects on human health and on the environment as a whole'. Member states have responsibility for implementing the directive and for the accurate assessment of air quality.

In a similar way, water quality is carefully monitored and regulated under EU environmental legislation. EU water quality standards are laid down in a range of areas, including provision of drinking water, water for fish and shellfish, and for bathing beaches. Values are laid down which limit emission of nitrates, of urban waste water, sewage sludge and pesticides. The 'Blue Flag' scheme for beaches used by holidaymakers is an example of the application of these standards. Blue Flags may be awarded annually to bathing beaches, providing they comply with certain microbiological standards and satisfy a range of other criteria, including their amenities. In 1992, 17 beaches in the UK qualified for the award of a Blue Flag. Part of the requirement is that information relating to quality of water on such bathing beaches must be made available to the public.

Through its legislation the EU issues official instructions, known as 'Directives'. Two such Directives are referred to here, just as examples, though the details are too extensive to cover more fully.

The Nitrates Directive (Council Directive 91/676/EEC, in 1991) aims to protect fresh, coastal and marine water against nitrate pollution. More specifically it covers

- surface waters (particularly those that may be used to supply drinking water)
- groundwaters
- bodies of water, such as freshwater lakes, estuaries and coastal waters, especially those likely to become eutrophic.

In the surface and groundwater there is particular concern if nitrate concentration exceeds 50 mg dm^{-3} (or is likely to do so). The Directive provides for the identification of 'Nitrate Vulnerable Zones'. This, for example, could be an area liable to suffer from leaching of nitrate from farms, derived from manures or fertilisers. Certain restrictions are imposed on farmers in such areas to control applications of nitrogen fertilisers and manure on their land.

The Bathing Waters Directive aims to maintain a standard of bathing waters for the protection of human health and to ensure appropriate standards are maintained with regard to the amenity value of the area. A revised Directive for the Quality of Bathing Water was proposed in October 2002 (COM[2002]581). The areas covered can be coastal or inland waterways, including beaches that are popular with visitors and where bathing is either authorised or not prohibited. Waters covered by the Directive are monitored regularly during the bathing season. In the UK, this is from late May or early June, through to mid- or late September. A number of factors are measured to ensure they are within the acceptable standards. The factors include levels of certain microorganisms, pH, presence of oils and detergents and colour and clarity of the water.

To be effective, controls (of pollutants) must be enforced and these must operate at the personal and industrial company levels as well as at a national level. Enforcement is frequently through financial penalties for failures. There are signs, however, of increasing use of positive encouragement for good practice, through tax incentives and through education. The overall hope is that, through modification of personal behaviour and cooperation at an international level, there can be a noticeable reversal of the adverse effects on the environment of the pollution that is generated by human (anthropogenic) activities.

EXTENSION MATERIAL

The River Thames – an example of recovery from pollution

We can illustrate this by the effects of pollution in the River Thames over the last 200 years. The River Thames was once known as a notable salmon river, but by about 1830 these fish had disappeared because of the level of pollution preventing them from reaching their spawning grounds upstream. Many other species disappeared during the hundred years or more leading up to the 1950s, but have gradually reappeared after massive clean-up operations from the mid-1960s. Some examples, illustrated in Figure 9.23, indicate the success of the measures taken. The key factors in bringing about this improvement have been introduction of measures to control discharge of effluents into the river and the recognition of the importance of maintaining an adequate level of dissolved oxygen in the water.

EXTENSION MATERIAL

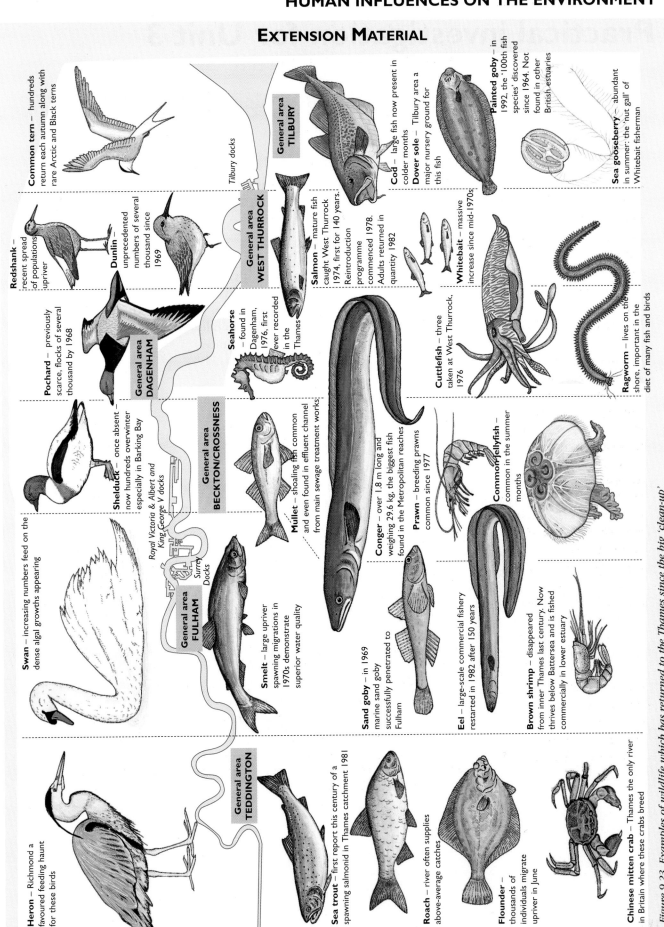

Common tern – hundreds return each autumn along with rare Arctic and Black terns

Redshank – recent spread of populations upriver

Dunlin – unprecedented numbers of several thousand since 1969

Tilbury docks

General area TILBURY

General area WEST THURROCK

Cod – large fish now present in colder months

Dover sole – Tilbury area a major nursery ground for this fish

Painted goby – in 1992, the '100th fish species' discovered since 1964. Not found in other British estuaries

Sea gooseberry – abundant in summer: the 'nut gall' of Whitebait fisherman

Salmon – mature fish caught West Thurrock 1974, first for 140 years. Reintroduction programme commenced 1978. Adults returned in quantity 1982

Whitebait – massive increase since mid-1970s

Seahorse – found in Dagenham, 1976, first ever recorded in the Thames

Cuttlefish – three taken at West Thurrock, 1976

Ragworm – lives on the shore, important in the diet of many fish and birds

Pochard – previously scarce, flocks of several thousand by 1968

General area DAGENHAM

Shelduck – once absent now hundreds overwinter especially in Barking Bay

Royal Victoria & Albert and King George V docks

General area BECKTON/CROSSNESS

Mullet – shoaling fish common and even found in effluent channel from main sewage treatment works

Conger – over 1.8 m long and weighing 29.6 kg, the biggest fish found in the Metropolitan reaches

Prawn – breeding prawns common since 1977

Common jellyfish – common in the summer months

Swan – increasing numbers feed on the dense algal growths appearing

General area FULHAM

Surrey Docks

Smelt – large upriver spawning migrations in 1970s demonstrate superior water quality

Sand goby – in 1969 marine sand goby successfully penetrated to Fulham

Eel – large-scale commercial fishery restarted in 1982 after 150 years

Brown shrimp – disappeared from inner Thames last century. Now thrives below Battersea and is fished commercially in lower estuary

General area TEDDINGTON

Heron – Richmond a favoured feeding haunt for these birds

Sea trout – first report this century of a spawning salmonid in Thames catchment 1981

Roach – river often supplies above-average catches

Flounder – thousands of individuals migrate upriver in June

Chinese mitten crab – Thames the only river in Britain where these crabs breed

Figure 9.23 Examples of wildlife which has returned to the Thames since the big 'clean-up'

Practical investigation for Unit 3

Estimation of pyramids of number and of biomass

Introduction

The purpose of this practical is to obtain data which can be used to construct pyramids of number or of fresh biomass. The ecological community chosen will, of course, depend on accessibility, but the exercise could be carried out on a rocky shore, in woodland or in open grassland.

Materials

- 1 m² or 0.5 m² quadrat
- Trowel
- Scissors
- Large white sorting tray
- Hand lens

- Beakers
- Forceps
- Pooter
- Identification key
- Balance

Method

1 Select an area and place your quadrat carefully. On a rocky shore you should ensure that the entire quadrat is occupied by only one type of community.
2 Collect the leaf litter, or cut plants at the base, and place in the white tray. If appropriate, record the number of individual plants.
3 Search carefully and remove all the animals present. Smaller animals may be removed with a pooter (Figure P.8), larger animals should be handled with forceps, or fingers. Place in suitable containers, such as plastic beakers.
4 Weigh the plant material.
5 Sort the animals into two groups: primary consumers (herbivores) and secondary consumers (carnivores).
6 Weigh the groups of animals separately and record the total number of animals in each group.
7 Return the animals to their habitat.

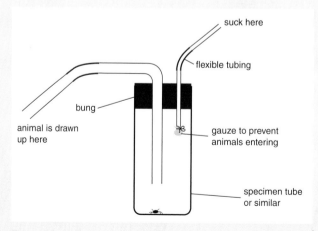

Figure P.8 A pooter

Results and discussion

1 Construct pyramids of number and of biomass. A horizontal scale is chosen to represent either the numbers of organisms present per m^2 or the biomass per m^2. The lower block represents the producers (plants), the middle block the primary consumers (herbivores) and the upper block the secondary consumers (carnivores).

2 Consider the advantages and disadvantages of using pyramids of numbers and biomass to represent an ecosystem.

Further work

1 You could use this method to compare two communities, such as different areas of leaf litter. It is important that comparable samples are used in each case.

2 How could you adapt the method to construct a pyramid of number and of biomass in a pond?

Assessment questions

The following questions have been chosen from recent Unit tests on the content of Units 2B, 2H and 3 of the Edexcel Biology and Biology (Human) Advanced Subsidiary GCE specification. The style and format of these questions will be similar to those in future tests. The shorter, structured questions are designed to test mainly knowledge and understanding of the topics, and the longer questions contain sections in which you may be required to demonstrate skills of interpretation and the evaluation of data. Some sections of the longer questions may require extended answers, for 4 or more marks, but there will no longer be a requirement for a 'free prose' answer worth 10 marks.

As some topics in these Units are interlinked, you may find that some questions require knowledge of more than one section of the specification.

Questions set specifically on the content of Unit 2H, Biology (Human) have been included and are identified.

The content of Unit 3 is common to both Biology and Biology (Human).

Chapter 1

1 The table below lists some enzymes associated with carbohydrate digestion in humans, their site of secretion and product(s) of their action.

Copy and complete the table by filling in the blank spaces.

Enzyme	Site of secretion	Product(s)
Amylase		
	Lining (mucosa) of ileum	Glucose and galactose
Sucrase		Glucose and fructose

(Total 4 marks)
(Edexcel GCE Biology 6102/2B, January 2003)

2 The diagram below shows part of the lower surface of a leaf, as seen using a light microscope.

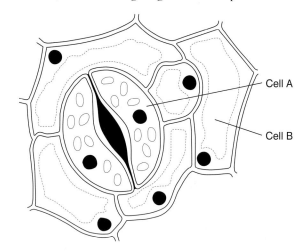

(a) Name the cells labelled A and B. **[2]**

(b) In an investigation, the concentrations of potassium ions (K^+) were measured in cells A and B, when the stoma was open and when it was closed. The results are shown in the table below.

Cell	Concentraton of potassium ions/mol dm⁻³	
	Stoma open	**Stoma closed**
A	0.45	0.10
B	0.10	0.20

(i) Describe these changes in potassium ion concentrations. **[2]**

(ii) Suggest how changes in potassium ion concentrations are involved in the mechanism for opening of stomata. **[3]**

(Total 7 marks)
(Edexcel GCE Biology 6102/2B, June 2002)

3 The diagram shows the structure of part of the ileum as seen in transvere section.

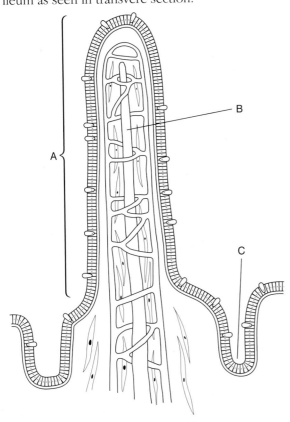

(a) Name the parts labelled A, B and C. **[3]**

(b) Describe **two** ways in which the structure of part A is adapted for the absorption of the products of digestion. **[4]**

(c) The table below lists some enzymes associated with carbohydrate digestion, their site of secretion and the products of their action.

Copy and complete the table by filling in the blank spaces.

Enzyme	Site of secretion	Products
	Pancreas	Maltose
Lactase		
Sucrase	Lining (mucosa) of ileum	

[4]

(Total 11 marks)

(Edexcel GCE Biology 6102/2B, June 2001)

4 The photograph below shows part of a tranverse section of a leaf, as seen using a light microscope.

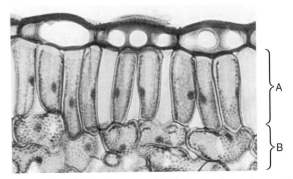

(a) Name the tissues labelled **A** and **B**. **[2]**

(b) Explain how the tissue labelled **B** is adapted for the function of gas exchange. **[3]**

(c) In an investigation into gas exchange in leaves, a maize leaf was placed in a dark chamber, and the mass of carbon dioxide released from the leaf in one hour was determined. The surface area of the leaf was also measured.

The results are shown in the table below.

Mass of carbon dioxide produced per hour/mg	Surface area of leaf/cm^2
4.076	29

(i) Calculate the mass of carbon dioxide released in one hour per unit area of leaf. Show your working. **[2]**

(ii) Suggest how the results would have differed if the leaf had been illuminated. **[2]**

(Total 9 marks)

(Edexcel GCE Biology 6102/2B, January 2003)

5 The table below refers to three different types of epithelia. Copy and complete the table by writing the name of each type of epithelium and giving **one** location of each.

Epithelium	Name	**One** location

(Total 6 marks)

(Edexcel GCE Biology 6112/2H, January 2003)

6 Read through the following passage on alveoli and gas exchange, then copy and complete it using the most appropriate word or words to fill the blanks.

During inspiration, air reaches the alveoli through small tubes known as . . . . The alveolar wall is composed of a . . . . layer of epithelial cells which . . . . the diffusion rate of oxygen from the alveolar air into the . . . . surrounding the alveoli.

(Total 4 marks)

(Edexcel GCE Biology 6102/2B, June 2003)

Chapter 2

1 The photograph below shows a transverse section of part of a root, as seen using a light microscope.

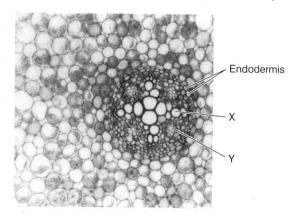

(a) Which label, **X** or **Y**, is a xylem vessel? **[1]**

(b) Describe the role of the endodermis. **[3]**

(c) In an investigation, the rate of uptake of water by a sunflower was measured at intervals of two hours from 08.00 hours until 06.00 hours on the next day. The results are shown in the table below

Time (24 hour clock)	Rate of uptake of water/g per 2 hours
08.00	3
10.00	12
12.00	29
14.00	38
16.00	39
18.00	40
20.00	12
22.00	9
24.00	8
02.00	5
04.00	2
06.00	2

(i) Describe the changes in the uptake of water which occurred during this investigation. **[2]**

(ii) Suggest an explanation for the change in the uptake of water which occurred between 08.00 hours and 14.00 hours. **[2]**

(Total 8 marks)

(Edexcel GCE Biology 6102/2B, January 2003)

2 Experiments were carried out to investigate the uptake of mineral ions by barley roots.

In the first investigation, isolated barley roots were immersed in an aerated culture solution containing potassium ions (K^+) and nitrate ions (NO_3^-). After ten hours, the roots were removed and the concentrations of these ions in the cell sap were determined. The results are shown in the table.

Ion	Concentration in culture solution/ mmol per dm³	Concentration in cell sap/ mmol per dm³
Potassium	7.98	97.8
Nitrate	7.29	38.1

(a) Suggest why the culture solution was aerated **[2]**

(b) These results show that the concentration of potassium ions in the cell sap is 12.3 times greater than that in the culture solution. This is referred to as the **accumulation ratio**.

Calculate the accumulation ratio for nitrate ions. Show your working. **[2]**

(c) What do these results suggest about the mechanism for the uptake of potassium and nitrate ions? Explain your answer. **[2]**

(d) In a further experiment, the effect of temperature on the uptake of potassium ions was investigated. Isolated barley roots were kept in aerated nutrient solutions at a range of temperatures, and the concentrations of potassium ions in the cell sap were measured after ten hours. The results are shown in the table below.

Temperature / °C	Concentration of potassium ions in cell sap/mmol per dm³
6	35
12	42
18	70
24	95
30	110

(i) What effect does temperature have on the concentration of potassium ions in the cell sap?

(ii) Suggest an explanation for these results. **[4]**

(Total 10 marks)

(Edexcel GCE Biology 6102/2B, June 2001)

3 *(a)* Cardiac muscle contracts myogenically. Explain what is meant by the term **myogenic**. **[2]**

(b) The diagram below shows structures in the heart which are concerned with the coordination of contraction.

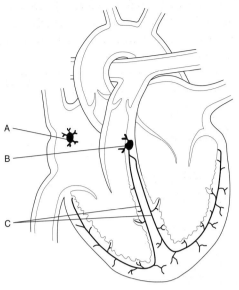

(i) Name parts **A**, **B** and **C**. **[3]**

(ii) Explain how the structures shown in the diagram coordinate the contraction of the heart. **[3]**

(Total 8 marks)

(Edexcel GCE Biology 6102/2B, January 2003)

4 The diagram below shows the formation of tissue fluid from part of a capillary.

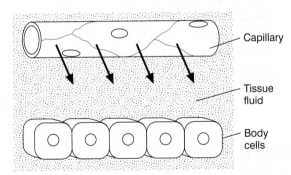

(a) Describe how tissue fluid is formed. **[2]**

(b) The table below shows the concentration of some solutes in blood plasma and tissue fluid.

Solute	Concentration in blood plasma / mmol dm³	Concentration in tissue fluid / mmol dm³
Potassium ions	4.0	4.0
Sulphate ions	0.5	0.5
Protein	2.0	Less than 0.1

(i) Compare the concentrations of these solutes in blood plasma and tissue fluid. **[2]**

(ii) Suggest explanations for the difference in the concentration of these solutes in blood plasma and tissue fluid **[3]**

(Total 7 marks)
(Edexcel GCE Biology 6102/2B, January 2003)

5 The graph below shows oxygen dissociation curves for human haemoglobin and myoglobin.

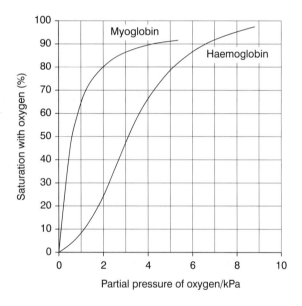

(a) From the graph, find the partial pressures of oxygen at which myoglobin and haemoglobin are 50% saturated with oxygen. **[1]**

(b) Describe the role of myoglobin. **[3]**

(c) On the graph, draw a curve to show the dissociation curve for fetal haemoglobin. **[2]**

(d) If the partial pressure of carbon dioxide increases, the dissociation curve for haemoglobin moves to the right.

(i) What name is given to this effect? **[1]**

(ii) Explain the importance of this effect. **[3]**

(Total 10 marks)
(Edexcel GCE Biology 6102/2B, June 2003)

Chapter 3

1 The photomicrographs show a transverse section through a leaf of *Ammophila*, which is a xerophyte. The large photomicrograph shows details of the tissues inside the box.

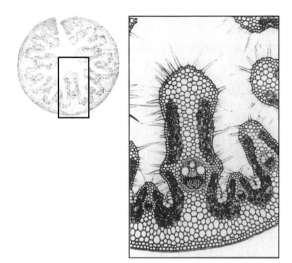

Describe **three** ways in which this leaf is adapted to reduce water loss.

(Total 6 marks)
(Edexcel GCE Biology 6102/2B, June 2001)

2 A freshwater stream was sampled over a distance of 4.0 km to determine the abundance of an aquatic invertebrate. The oxygen concentration of the water was measured over the same distance. The results are shown in the graphs below.

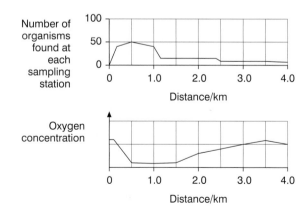

(a) Name **one** aquatic invertebrate that might show this distribution along the stream. **[1]**

(b) Suggest **two** adaptations that would enable an invertebrate to thrive between 0.2 and 1.0 km along the stream. In each case state how the adaptation assists survival. **[4]**

(Total 5 marks)
(Edexcel GCE Biology 6102/2B, June 2001)

3 The diagrams below show two invertebrates found in freshwater habitats. Rat-tailed maggots are found in stagnant or slow moving water. Stonefly nymphs live in fast flowing streams.

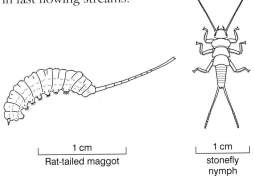

1 cm	1 cm
Rat-tailed maggot	stonefly nymph

(a) Suggest how rat-tailed maggots are adapted to their habitat. **[4]**

(b) Suggest what adaptations stonefly nymphs might have for living in fast flowing water. **[2]**

(Total 6 marks)
(Edexcel GCE Biology 6102/2B, June 2003)

4 The photomicrograph below shows part of a transverse section through the stem of *Juncus*, a hydrophyte (a plant that lives in water).

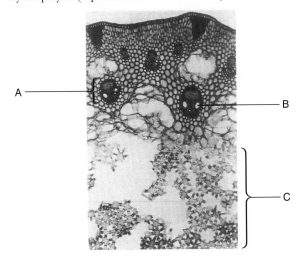

(a) Name the parts labelled A and B. **[2]**

(b) Suggest **two** roles for the tissue labelled C. **[2]**

(c) Some hydrophytes have finely divided, feathery submerged leaves and spreading leaves floating on the surface of the water. Explain how each of these features is an adaptation to the environment in which hydrophytes live.

Finely divided, feathery submerged leaves

Spreading leaves floating on the surface of the water **[4]**

(Total 8 marks)
(Edexcel GCE Biology 6102/2B, January 2002)

Chapter 4

1 When humans ascend rapidly to high altitude, for example in a balloon flight, they may become dizzy, feel sick and eventually lose consciousness. The graph below shows the relationship between altitude and the time taken to lose consciousness.

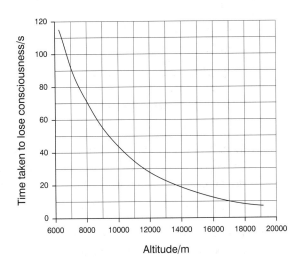

(a) From the graph, find the time taken to lose consciousness at 10 000 m. **[1]**

(b) Describe how **two** environmental conditions change during an ascent from sea level to 10 000 m. **[4]**

(c) Suggest reasons why mountaineers do not usually lose consciousness at the top of Mount Everest (approximately 9000 m). **[3]**

(Total 8 marks)
(Edexcel GCE Biology 6112/2H, June 2002)

ASSESSMENT QUESTIONS

2 *(a)* In humans, normal body temperature is maintained by a number of different mechanisms. If body temperature changes, information is sent to the temperature regulating centre, which then makes the appropriate response to bring body temperature back to normal.

In some extreme situations, if the temperature-regulating mechanisms fail, body temperature may fall below 35 °C.

(i) Name the part of the brain in which the temperature-regulating centre is situated. **[1]**

(ii) What term is used to describe the condition in which body temperature falls below 35 °C? **[1]**

(b) In an investigation into body temperature, the oral (mouth) temperature of a healthy person was measured every five minutes for 40 minutes. From five minutes after the start of the investigation until 20 minutes, the person was lying in a hot bath, at a temperature of 42 °C. After 20 minutes the person got out of the bath and sat on a chair.

The results of this investigation are shown in the table below.

Time/min	Oral temperature/ °C
0 (start of investigation	37.0
5	37.0
10	37.2
15	37.8
20	37.9
25	37.7
30	37.5
35	37.2
40	37.0

(i) Describe the changes in oral temperature which occurred during the investigation. **[3]**

(ii) Suggest explanations for the changes in oral temperature which occurred between the following time intervals: 0 to 20 minutes; 20 to 40 minutes **[5]**

(Total 10 marks)
(Edexcel GCE Biology 6112/2H, June 2002)

3 The graph below shows the oxygen dissociation curves for the haemoglobin from a human and from a llama. A llama is a mammal that lives in the high Andes of South America, often at altitudes above 5000 metres.

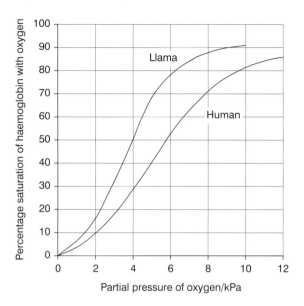

(a) From the graph, find the partial pressure of oxygen at which the haemoglobin of a llama and of a human is 50% saturated with oxygen. **[2]**

(b) Explain how the oxygen dissociation curve for the llama suggests that it is adapted for life at high altitudes. **[4]**

(c) On the graph, sketch a curve to show the effect of an increase in the partial pressure of carbon dioxide (the Bohr effect) on the dissociation curve of human haemoglobin. **[2]**

(d) Mountaineers may acclimatise to the conditions at high altitude. State what happens to the red blood cells after a period of time at high altitude and explain the importance of this change. **[2]**

(e) Mountaineers experience very low temperatures at high altitude. Describe **one** physiological effect of exposure to these low temperatures. **[2]**

(Total 12 marks)
(Edexcel GCE Biology 6112/2H, January 2003)

4 An investigation was carried out into the effect on the body temperature of naked humans of exposure to low environmental temperature. Two groups of adult males were studied, Europeans from a temperate climate and Australian Aboriginals from a climate with very hot days and cold nights. Both groups were exposed to an air temperature of 5 °C during a night of eight hours.

The results are shown in the table below.

Time/hours	Body temperature/ °C	
	Europeans	**Australian Aboriginals**
0	37.2	37.2
1	36.4	36.9
2	36.1	36.5
3	36.2	36.3
4	36.1	36.0
5	36.1	36.0
6	36.2	35.8
7	36.3	35.4
8	36.4	35.2

(a) Describe the changes in body temperature in these two groups. **[3]**

(b) Suggest how these results show that Australian Aboriginals are adapted to the climate in which they live. **[2]**

(Total 5 marks)
(Edexcel GCE Biology 6112/2H, June 2003)

Chapter 5

1 *(a)* Explain what is meant by the term **pollination**. **[2]**

(b) The diagram shows the structure of a grass flower.

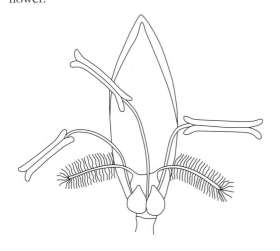

Describe **two** ways in which this flower is adapted for wind pollination. **[4]**

(c) This grass flower can be self-pollinated. Suggest how the flowers of other grasses might be adapted to avoid self pollination. **[3]**

(Total 9 marks)
(Edexcel GCE Biology 6102/2B, June 2001)

2 The diagram below shows a germinating pollen grain as seen using a light microscope.

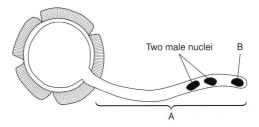

(a) Name the parts labelled A and B. **[2]**

(b) Describe the roles of the two male nuclei during fertilisation. **[4]**

(c) An investigation was carried out into the effect of sucrose concentration on the germination of pollen grains from two species of plants, *Bauhinia purpurea* and *Camellia japonica*. The results are shown in the graph below.

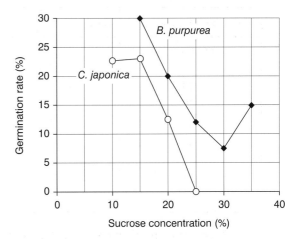

(i) What is the optimum concentration of sucrose for the germination of pollen grains from both species **[1]**

(ii) Compare the germination rate of these two species as the concentration of sucrose increases from 20%. **[3]**

(Total 10 marks)
(Edexcel GCE Biology 6102/2B, June 2002)

ASSESSMENT QUESTIONS

3 The graph shows changes in the concentration of progesterone in the blood during the menstrual cycle.

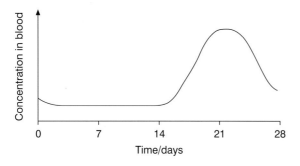

(a) Sketch the graph and draw a line to show how the concentration of **oestrogen** changes from day 0 to day 28 of the menstrual cycle. **[2]**

(b) State where progesterone is produced during the menstrual cycle. **[1]**

(c) State **one** effect of progesterone. **[1]**

(d) Explain why the concentration of progesterone decreases towards the end of the menstrual cycle. **[2]**

(e) Explain what happens to the concentration of progesterone if fertilisation and implantation occur. **[2]**

(Total 8 marks)
(Edexcel GCE Biology 6102/2B, June 2001)

4 The flow chart below shows the sequence in which some cells are formed during spermatogenesis in the mammalian testes.

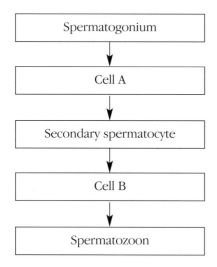

(a) State the part of the testis where spermatogenesis occurs. **[1]**

(b) Name cells A and B. **[2]**

(c) Explain the importance of meiosis in the formation of spermatozoa. **[3]**

(Total 6 marks)
(Edexcel GCE Biology 6102/2B, June 2003)

5 (a) Explain what is meant by **implantation** in relation to human reproduction. **[2]**

(b) The diagram below shows the structure of part of a human placenta and umbilical cord.

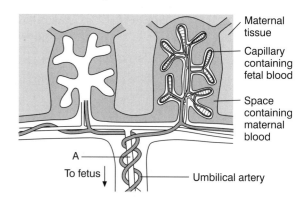

(i) Name the part labelled **A**. **[1]**

(ii) Name **two** substances which would be present in a higher concentration in the umbilical artery than in the mother's blood. **[2]**

(iii) With reference to the diagram, suggest how the structure of the placenta enhances the transfer of substances between the blood of the fetus and the blood of the mother. **[3]**

(iv) Shortly after the birth of the baby, the placenta leaves the uterus as the afterbirth, as a result of continued contraction of the uterine muscles. Name the hormone that causes this contraction. **[1]**

(Total 9 marks)
(Edexcel GCE Biology 6102/2B, January 2003)

6 Explain what is meant by each of the following terms.

(a) Growth **[3]**

(b) Menopause **[3]**

(Total 6 marks)
(Edexcel GCE Biology 6112/2H, January 2003)

Chapter 6

1 The diagrams below show the skull of a sheep (a herbivore) and a dog (a carnivore).

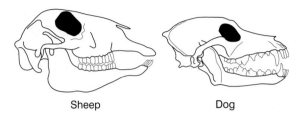

Sheep   Dog

Describe **three visible** differences between the teeth of the sheep and the dog. For each difference, explain how it is related to the differences between their diets.

(Total 6 marks)

(Edexcel GCE Biology 6103/03, January 2003)

2 The diagram shows part of the root system of a pea plant. Pea plants are members of the family Papilionaceae.

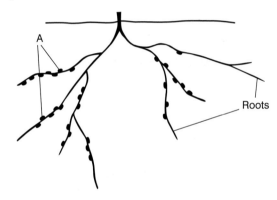

(a) Name the structures labelled A. **[1]**

(b) Name the bacterium that would be found within the cells of these structures. **[1]**

(c) State the term used to describe the type of association between members of the Papilionaceae, such as pea plants, and this bacterium. **[1]**

(d) Describe the part played by this bacterium in the nitrogen cycle. **[3]**

(Total 6 marks)

(Edexcel GCE Biology 6103/03, June 2002)

3 The diagram below shows part of the adult stage of the tapeworm, *Taenia*, which is an endoparasite of mammals.

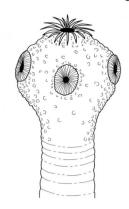

(a) State where the adult stage of *Taenia* would be found in the body of the host mammal. **[1]**

(b) Explain how the hooks and suckers shown in the diagram enable *Taenia* to be a successful endoparasite. **[2]**

(c) Give **two** features of *Taenia*, other than those shown in the diagram, that are adaptations to the parasitic mode of nutrition. **[2]**

(d) Explain how the mode of nutrition shown by a parasite, such as *Taenia*, differs from that shown by a fungus, such as *Rhizopus*. **[3]**

(Total 8 marks)

(Edexcel GCE Biology 6103/03, June 2003)

Chapters 7, 8 and 9

1 The diagram shows a pyramid of numbers, which is a diagrammatic way of representing the feeding relationship within an ecosystem.

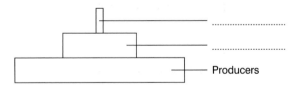

Producers

(a) Complete the diagram by writing the names of the second and third trophic levels in the spaces provided. **[2]**

(b) Describe how you would collect data to enable a pyramid of numbers to be drawn. **[4]**

(c) Explain why the number of individuals usually decreases along a food chain. **[2]**

(Total 8 marks)

(Edexcel GCE Biology 6103/03, January 2002)

ASSESSMENT QUESTIONS

2 Scientists estimate that the atmosphere holds 755 gigatonnes (Gt) of carbon, mostly as carbon dioxide. This value is increasing each year as a result of human activities, such as the burning of fossil fuels and of timber. The diagram below shows a simplified 'balance sheet' of the carbon cycle for one year.

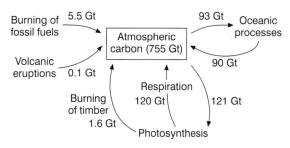

(a) Calculate the increase in atmospheric carbon during this year. Show your working. **[3]**

(b) Describe and explain how carbon can be removed from the terrestrial (land based) part of the cycle for long lengths of time. **[4]**

(c) By reference to the information in the cycle, suggest reasons for the net effect of oceanic processes on the carbon content of the atmosphere. **[4]**

(d) Suggest how the quantities of carbon dioxide released by the combustion of fossil fuels could be reduced. **[3]**

(Total 14 marks)
(Edexcel GCE Biology 6103/03, January 2002)

3 The removal of forest trees to supply wood, or so that land can be used for other purposes, has long been common practice. However, the last 250 years have seen an increase in the rate of deforestation.

One of the results of deforestation may be that there is a net increase of carbon dioxide in the atmosphere. During the period from 1958 to 1980, it is estimated that there was a net increase of 57.3×10^9 million tonnes of carbon in the atmosphere, mainly in the form of carbon dioxide. Much of this increase is thought to be a direct result of the clearance of the forests.

The photograph below was taken from space by satellite, showing an area of Amazonian rainforest partially cleared in 1998.

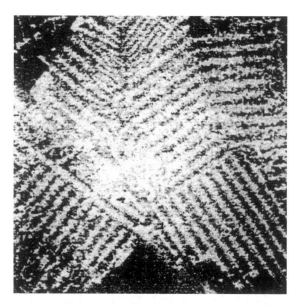

(a) State **two** ways in which the wood cleared from forests may be used. **[2]**

(b) Suggest, giving a reason, **one** possible use for the cleared land shown in the photograph. **[2]**

(c) Explain how the biodiversity in this area might be affected by deforestation. **[2]**

(d) Explain why extensive clearance of forests might lead to '*a net increase of carbon dioxide in the atmosphere*' (Paragraph 2) **[3]**

(e) With reference to a specific method of forestry management, explain what is meant by the sustainable management of forests. **[3]**

(Total 12 marks)
(Edexcel GCE Biology 6103/03, June 2001)

4 Study the passage and data below and then answer the questions that follow.

A study of river pollution in South Wales

The map in Figure 1 shows the course of a river that flows through a moderately-sized industrial town in South Wales. At the lower part of the town, the river flows through a tunnel that passes under a large tin plating works.

ASSESSMENT QUESTIONS

Figure 1

Map showing the course of the river through the town

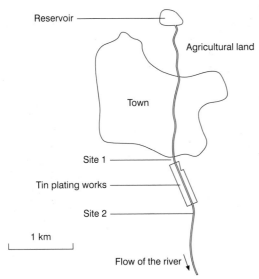

In the 19th century, the effluent (waste water) from the tin plating works was allowed to pass untreated into the river as it flowed through the tunnel. In more recent times, the effluent has been treated to reduce the level of pollutants entering the river.

A study was carried out to assess the water quality of the river. Two sites were chosen as shown in Figure 1. Site 1 was just above the entry to the tunnel. Site 2 was just below the point at which the river emerges above ground as it leaves the tunnel.

The degree of pollution was assessed by identifying the species of aquatic invertebrates that were present. The number of different species within each of 14 families was counted and recorded in the table shown in Figure 2.

Each family was given an index number according to the quality of the water in which it is usually found. This number was multiplied by the number of species to give a score for each family.

The scores for the two sites are shown in Figure 2. The degree of pollution and river quality was then assessed by reference to the table shown in Figure 3.

The invertebrates were collected using a kick sampling technique. This involves disturbing the rocks and debris at the bottom of the river to dislodge the invertebrates. They are then caught in a net that is held open downstream.

Figure 2

Data for the two sites

| Invertebrate group | Species index number | Site 1 | | Site 2 | |
		Number of species found	Group score	Number of species found	Group score
Ecdyonuridae	10	5	50	0	
Ephemeridae	10	2	20	0	
Perlidae	10	2	20	0	
Leuctridae	10	2	20	0	
Corixidae	5	2	10	0	
Dytiscidae	5	4	20	2	
Helodidae	5	3	15	2	
Tipulidae	5	6	30	5	
Baetidae	4	4	16	3	
Glossiphoniidae	3	0	0	2	
Erpobdellidae	3	2	6	3	
Limnaeidae	3	2	6	2	
Chironomidae	2	4	8	3	
Hydrobiidae	2	3	6	2	
		Total score for Site 1	227	Total score for site 2	

ASSESSMENT QUESTIONS

Figure 3

Scoring table for assessment of pollution

Total site score	Degree of pollution	River quality
0–100	Very severe	Very poor
101–200	Severe	Poor
201–300	Moderate	Moderate
301–400	Slight	Good
401–500	None	Excellent

(a) Complete the table in Figure 2 by calculating the scores for each invertebrate group at Site 2, and the total score for the site. **[2]**

(b) Using the information in Figure 2 and Figure 3, compare the results for Site 1 with those for Site 2. **[4]**

(c) Suggest **two** reasons why the results of kick sampling may be unreliable. **[2]**

(d) Suggest why aquatic invertebrates can be used as indicators of pollution. **[2]**

(e) Suggest and explain how the water may be affected as it flows past the agricultural land above the town. **[4]**

(f) Suggest why the concentration of dissolved phosphate in the water may increase as the river flows through the town. **[2]**

(Total 16 marks)
(Edexcel GCE Biology 6103/03, January 2002)

Mark schemes

In the mark schemes the following symbols are used:
; indicates separate marking points
/ indicates alternate marking points
eq. means correct equivalent points are accepted
{} indicate a list of alternatives

Unit 2

Chapter I

Enzyme	Site of secretion	Product(s)
	Pancreas / salivary glands ;	Maltose ;
Lactase ;		
	{Lining / epithelium / mucosa} of {small intestine / ileum / duodenum} ;	

(Total 4 marks)

2 (a) A Guard cell ;

B Epidermal cell / epidermis cell / subsidiary cell ;   **[2]**

(b) (i) In cell A the concentration (of potassium) is higher when the stoma is open / converse statement ;

In cell B the concentration (of potassium) is higher when the stoma is closed / converse statement ;

When stoma is closed concentration (of potassium) is lower in cell A than in cell B / converse statement ;

When stoma is open concentration (of potassium) is higher in cell A than in cell B / converse statement ;

Credit quantitative manipulated comparison e.g. difference calculated ;

[2]

(ii) Increased concentration of potassium ions reduces solute potential / eq ;

Water moves into (guard) cell ;

By osmosis / down water potential gradient / down diffusion gradient ; [accept diffusion gradient if "water" is understood]

Increases turgor / hydrostatic pressure, of guard cells / A / makes cells turgid (therefore stomata open) ;   **[3]**

(Total 7 marks)

3 (a) **A** – villus ; [accept villi]

B – lymph vessel / lacteal ; [accept lymph capillary]

C – crypt (of Lieberkuhn) ;   **[3]**

(b) (Villi / microvilli, provide) large surface area ;
To increase rate of absorption / uptake ;

Simple / single layered, epithelium / single layer of cells ;
Provides short (diffusion) pathway / distance ;

Presence of capillaries ;
To absorb / transport, amino acids / glucose / any other correctly named substance
or
To maintain, diffusion gradient / concentration gradient ;

Capillaries near surface ;
Provides short (diffusion) pathway / distance ;

Presence of lacteal ;
To absorb / transport, fatty acids / lipids / fats / oils / fat soluble vitamins ;

Presence of (smooth) muscle ;
To assist, contact with contents / mixing ;

[2+2]

(Max 4 marks)

MARK SCHEMES

(c)

Enzyme	Site of secretion	Product(s)
	mouth/salivary gland/ pancreas ;	maltose ;
Lactase ;		
	lining/epithelium/ mucosa, of small intestine/ileum/ duodenum ;	

[3]

(Total 11 marks)

4 (a) **A** Palisade (mesophyll / layer) ;

   B Spongy (mesophyll / layer) ;   [2]

   (b) Large surface area ;

   Thin (cell) walls ;

   Presence of air spaces ;

   Reference to increased rate of diffusion ;   [3]

   (c) (i) 4.076 ÷ 29 ;

   = 0.14 / 0.141 mg per cm^2 (per hour) ;   [2]

   (ii) Less carbon dioxide released / no CO_2 {released / produced} / CO_2 taken up rather than released ;

   Because photosynthesis occurs / carbon dioxide used in photosynthesis / eq ;   [2]

   (Total 9 marks)

5

Epithelium	Name	One location
	Cuboidal ;	Nephron ;
	Squamous / pavement ;	Alveolus / lining of mouth / Bowman's capsule / lining blood vessels ;
	Columnar ;	Ileum / small intestine ;

[The locations given above are the ones named in the specification, other correct locations can be accepted]

(Total 6 marks)

6 Bronchioles / alveolar ducts ;

Single / thin / squamous / flattened / type I pneumocytes ;

Increases / eq ;

Blood / capillaries / red blood cells ;

(Total 4 marks)

Chapter 2

1 (a) Xylem vessel correctly indicated as X ;   [1]

   (b) 1. Reference to Casparian strip / suberin / waxy layer ;

   2. Waterproof ;

   3. Water directed from apoplast pathway / eq ;

   4. Into symplast pathway / eq ;

   5. Reference to regulation of uptake / control of uptake ;

   6. Active transport of mineral ions ;

   7. Reference to passage cells ;   [3]

   (c) (i) Uptake increases (from 08.00) {to 18.00 / up to 14.00 then ± constant to 18.00} ;

   Decreases from 18.00 to {04.00 / 06.00} ;

   Credit manipulated figure reference ;   [2]

   (ii) Increase in transpiration / eq ;

   Increase in {temperature / wind speed / stomatal opening / light intensity} OR decrease in humidity ;   [2]

   (Total 8 marks)

2 (a) To provide oxygen ;

   For aerobic respiration ;

   Ref. requirement for ATP (produced by respiration) ;   [2]

   (b) 38.1 ÷ 7.29 ;

   = 5.2 ; [accept 5.23]   [2]

   (c) They are taken up, by active transport / active uptake / actively ;

   Because the concentration in cell sap is higher than concentration in culture solution / ions are taken up against a concentration gradient ;   [2]

(d) (i) Conc of ions increases as temperature increases ;

Credit any quantitative description ; **[2]**

(ii) (Increase in temperature) increases, kinetic energy / movement (of ions) ;

Increases enzyme activity ;

Increases, (rate of) respiration / production of ATP ;

Increases, (rate of) collisions of ions with transporter protein / eq ; **[2]**

(Total 10 marks)

3 *(a)* Spontaneous / automatic (contraction) / self-stimulating ;

No {nerve impulse / eq} required / no nerve stimulation ; **[2]**

(b) (i) A SAN / sino-atrial node / pacemaker ;

B AVN / atrio-ventricular node ;

C Bundle of His / Purkyne fibres / Purkinje tissue ; **[3]**

(ii) 1. {SAN / A / pacemaker} initiates impulse / eq ;

2. {Impulse / eq} to {AVN / B} then delay ;

3. So atrial {systole / contraction} before ventricular systole / ventricles fill with blood before systole ;

4. {Bundle of His / C} conducts impulse to (base of) ventricles ;

5. Which contract from {apex / eq} upwards ; **[3]**

(Total 8 marks)

4 *(a)* {Blood / hydrostatic} pressure ;

Forces water (and solutes) out (of capillary) ;

Reference to pores / permeability of capillary wall / eq / single cell layer of capillary wall / squamous epithelium ; **[2]**

(b) (i) Potassium and sulphate ions have the same concentration (in plasma and tissue fluid) ;

Higher concentration of protein in plasma (than in tissue fluid) / converse ; **[2]**

(ii) 1. Capillary (freely) permeable to ions / eq ;

2. (because) ions are small ;

3. Capillary (much) less permeable to proteins ;

4. (because) proteins large ;

5. Therefore most stay in {capillaries / plasma} / cannot pass through pores ; **[3]**

(Total 7 marks)

5 *(a)* (Myoglobin) 0.6 kPa and (haemoglobin) 3.1 kPa ; **[1]**

(b) (Myoglobin acts as) a store of oxygen ;

In muscle ;

Releases oxygen at (very) low partial pressure (of oxygen) ;

Provides oxygen during strenuous activity / allows muscle to respire aerobically for longer / eq ;

(Myoglobin) has a higher affinity for oxygen (than haemoglobin) ; **[3]**

(c) S-shaped curve, starting at 0,0 ;

Drawn to the left of the haemoglobin curve ; **[2]**

(d) (i) Bohr (effect / shift) ; **[1]**

(d) (ii) The affinity of haemoglobin (for oxygen) is reduced ;

Therefore releases oxygen more readily / percentage saturation with oxygen falls / dissociates more readily / more oxygen is released ;

At the same partial pressure of oxygen ;

Oxygen is released to {actively respiring / active / exercising} tissues / eq ; **[3]**

(Total 10 marks)

MARK SCHEMES

Chapter 3

1   Thick cuticle ;
    Impermeable to water *or*
    Reduces, transpiration / diffusion of water (vapour)
    / reduces evaporation ;

    Few stomata / stomata on inside of leaf ;
    Reduces, transpiration / diffusion of water (vapour)
    / reduces evaporation ;

    Leaf rolled / eq ;
    Reduces (exposed) surface area / stomata open
    into enclosed space / maintains high humidity
    inside leaf / reduces transpiration / reduces
    evaporation ;

    Stomata in pits / sunken stomata ;
    Reduce air movement / increase humidity / reduces
    transpiration / reduces evaporation ;

    Presence of hairs / eq ;
    Reduce air movement / trap moist air (next to leaf)
    / reduce diffusion (gradient) / reduces transpiration
    / reduces evaporation ;

    Presence of hinge cells ;
    To curl leaf ;   **[2+2+2]**

                                          (Total 6 marks)

2   *(a)* *Tubifex* / sludgeworm / rat-tailed larva / midge
           larva / *Chironomus* larva / bloodworm / gnat
           larva / mosquito larva   **[1]**

    (b) Haemoglobin / respiratory pigment (in blood) ;
           Picks up oxygen (at low external oxygen
           concentrations) / very high affinity for oxygen ;

           Very thin body wall ;
           Short oxygen (diffusion) pathway ;

           Blood (vessels) near body surface ;
           Short oxygen (diffusion) pathway ;

           Waving tail ;
           Circulates water next to body / eq. ;

           Reference to breathing tube / eq ;
           Goes to surface / to obtain oxygen from the
           air ;

           Presence of hairs / wing cases / elytra ;
           To trap an air bubble ;

           [Credit other correct example]   **[2+2]**

                                          (Max 4 marks)

                                          (Total 5 marks)

3   *(a)* Low oxygen concentration ;

           Reference to a breathing tube / eq ;

           To obtain oxygen / exchange oxygen and
           carbon dioxide ;

           From above the surface of the water ;

           Reference to hairs or lipids to prevent
           waterlogging of tube ;

           Reference to respiratory pigment / haemoglobin ;

           With a high affinity for oxygen ;   **[4]**

    (b) Reference to claws / hooks (on legs) ;

           To hold on to stones / eq ;

           Streamlined body shape / eq ;

           Less resistance to flowing water / can hide
           under stones ;   **[2]**

                                          (Total 6 marks)

4   *(a)* A   vascular bundle ;

           B   xylem ;   **[2]**

    (b) Storage of gases / air / oxygen / carbon dioxide ;

           Gas exchange / gas diffusion / gas movement ;

           Buoyancy / eq ;   **[2]**

    (c) *Submerged*:

           Less resistance to water flow ;

           Less likely to be damaged by water currents / eq ;

           Reference to gas exchange ;

           Large surface area ;

           Floating:

           Large surface area ;

           (Increased) exposure to air ;

           (Increased) light trapping ;

           Reference to photosynthesis ;

           Stomata on the upper surface ;   **[4]**

                                          (Total 8 marks)

Chapter 4

1 *(a)* 43 / 44 seconds ; **[1]**

(b) Air pressure / pO$_2$;
Decreases ;

Temperature ;
Decreases ;

Wind speed / force ;
Increases ;

Humidity ;
Decreases ;

UV / solar radiation ;
Increases ; **[4]**

(c) Reference to acclimatisation ;

Increased numbers of red blood cells ;

Increased haemoglobin ;

Increased oxygen carrying capacity of blood ;

Use of oxygen cylinders ;

Hyperventilation ;

Increased pulmonary capillaries / increased
pulmonary diffusion capacity ; **[3]**

(Total 8 marks)

2 *(a)* (i) Hypothalamus ; **[1]**

(a) (ii) Hypothermia ; **[1]**

(b) (i) Temperature stayed the same from 0 to
5 minutes ;
Increased from 5 to 20 minutes / peaked
at 20 minutes ;
Then tempeature decreased from 20 to
40 minutes ;
Same at the end as at start ;
Quantitative description (e.g. temperature
increased by 0.9 °C from 0 to 20 minutes ; **[3]**

(b) (ii) **0 to 20 minutes**

Temperature of water is higher than body
temperature ;
Heat gained by conduction ;
Therefore body temperature increased ;
Sweating is ineffective as a way of
reducing body temperature in water ;

20 to 40 minutes

Reference to vasodilatation ;
Increased heat loss by radiation ;
Increased sweating ;
Reference to evaporation (of water) ;
Reference to latent heat ;
Therefore body temperature decreased /
body cools ; **[5]**

(Total 10 marks)

3 *(a)* (Llama) 4 kPa ;

(Human) 5.8 kPa ; **[2]**

(b) Dissociation curve is to the left (of human) ;
Haemoglobin has a high affinity for oxygen ;
Reference to low pO$_2$ at high altitudes ;
Therefore saturates with oxygen / picks up
oxygen at low pO$_2$; **[4]**

(c) S-shaped curve starting close to zero ;
To the right of the human curve ; **[2]**

(d) Number of red blood cells increases /
haemoglobin content increases ;
Increases oxygen carrying capacity of blood / eq ;
[2]

(e) Vasoconstriction ;
Of superficial blood vessels ;
For heat conservation ;

OR

Reduced sweating ;
For water conservation ;
For heat conservation ;

OR

Increase in thyroxin production / increase in
shivering ;
Increases metabolic rate ;
Increases heat production ; **[2]**

(Total 12 marks)

4 *(a)* Europeans show the greatest / fastest drop in
the first hour (or converse) ;

Aboriginal show a steady / eq drop throughout
8 hours ;

Europeans' body temperature more or less
constant from 1 hour / fluctuates ;

Credit a manipulated, quantitative comparison ;
[3]

(b) Aboriginals are tolerant of / adapted to cold nights ;

They allow their body temperature to fall ;

No shivering ;

Marked vasoconstriction ;

Reduces energy requirements / less food required ;

Reference to starting the day with a lower body temperature to cope with heat during the day ; **[2]**

(Total 5 marks)

Chapter 5

1 (a) Transfer of pollen (grains) ;

From anther to stigma ;

By named agent, e.g. wind / water / insect / animal ; **[Max 2 marks]**

(b) Large anthers ;
Large numbers of pollen (grains) produced ;

Exposed / swinging / versatile, anthers / long filaments / exposed stamens ;
Pollen shed into air / shaken by wind ;

Exposed / feathery, stigmas ;
Large surface area to, catch / trap, pollen (grains from the air) ; **[2+2]**
[Max 4 marks]

(c) Protandry / male eq parts ripen before female eq ;

Protogyny / female eq parts ripen before male eq ;

Dioecious plants / individual plants either male or female ; **[Max 3 marks]**

(Total 9 marks)

2 (a) A   Pollen tube ;

B   Tube nucleus ; **[2]**

(b) One, fuses / fertilises / combines, with female nucleus / egg cell / eq ;

To form zygote ;

Which is diploid ;

One fuses with polar nuclei / fusion nucleus / primary endosperm nucleus ;

To form endosperm (nucleus) ;

Which is triploid / 3n ; **[4]**

(c) (i)   15% ; **[1]**

(ii) (Germination rate of) both decreases ;

(Germination rate of) *Bauhinia* always greater than that of *Camellia* / converse ; [greater at all concentrations must be implied]

(Germination rate of) *Camellia* reaches zero at 25%, *Bauhinia* minimum at 30% / does not reach zero / eq ;

Above 30% / this (germination rate of) *Bauhinia* (starts to) increase (*Camellia* stays at zero) ; **[3]**

(Total 10 marks)

3 (a) One peak before day 14 ;

One peak after day 14 ; **[2]**

(b) Corpus luteum / yellow body / ovary ; **[1]**

(c) Maintenance of, endometrium / lining of uterus / secretory phase / inhibits, LH / FSH secretion / inhibits the, LH / FSH releasing hormone freq ; **[1]**

(d) Regression / degeneration / eq, of corpus luteum / yellow body ;

Because less LH present / secretion of LH inhibited / inhibition of LH releasing factor ; **[2]**

(e) Remains high / eq ;

Because, corpus luteum / yellow body, persists / eq ;

Correct ref. to human chorionic gonadotrophin / HCG ; **[Max 2 marks]**

(Total 8 marks)

4 (a) Seminiferous tubule ; **[1]**

(b) Cell A   Primary spermatocyte ;

Cell B   Spermatid ; **[2]**

(c) Meiosis reduces the chromosome number / from haploid to diploid / 2n to n / eq ;

So the diploid number is restored at fertilisation / eq ;

(Meiosis results in) genetic variation ;

Four sperm / four gametes produced ; **[3]**

(Total 6 marks)

5 (a) Embedding / eq of blastocyst / embryo ;

In the endometrium / in the lining of the uterus ; **[2]**

(b) (i) Umbilical vein ; **[1]**

(b) (ii) Carbon dioxide ;

Urea ;

Fetal haemoglobin ; **[2]**

(b) (iii) Large surface area for increased diffusion rate / eq ;

Very vascular / many capillaries / counter current blood flow, so no static build up of substances / maintains diffusion gradient ;

Short distance between maternal and fetal blood for increased transfer ;

Maternal blood space / eq causes slower blood flow so more time for transfer ; **[3]**

(b) (iv) Oxytocin ; **[1]**

(Total 9 marks)

6 (a) Increase in body mass / length / size / height ;

Irreversible / permanent ;

By cell division / enlargement / mitosis / cell multiplication ;

Reference to increase in complexity / influence of hormones / named example / reference to allometric growth ; **[3]**

(b) Decrease in fertility / ovulation stops / menstruation stops / sperm production decreases ;

(Associated with) decreasing oestrogen / decreasing testosterone ;

Occurs in males and females ;

Occurs at approximately age 50 in females / 60 to 65 in males ;

Any associated feature, such as hot flushes / night sweats / lower sex drive / mood swings ; **[3]**

(Total 6 marks)

Chapter 6

1 Incisors in sheep {only on lower jaw / absent on upper jaw} / reference to the horny pad on the upper jaw
AND
Incisors in dog present on upper and lower jaw ;

Linked to {chopping / cropping / cutting / tearing} in sheep OR {gripping / nibbling} in dog ;

Canines {small / only in lower jaw / absent} in sheep
AND
Canines {large / pointed} in dog ;

Linked to {gripping / piercing / stabbing} in dog ;

Sheep has a diastema OR {gap / space} between {canines / incisors} and premolars / position clearly described
AND
Dog has no {diastema / gap} ;

Linked to {manipulation / eq} of food (by tongue) in sheep ;

{Premolars / molars / cheek teeth} of sheep {ridged / interlocking / not sharp}
AND
{Premolars / molars / carnassials} of dog {pointed / scissor blades / sharp / not ridged / not interlocking} ;

Linked to grinding in sheep OR {slicing / shearing / cutting / crushing bone} in dog ; **[3 × 2]**

(Total 6 marks)

2 (a) (root) nodules ; **[1]**

(b) *Rhizobium* ; **[1]**

(c) Mutualism / mutualistic / symbiosis / symbiotic ; **[1]**

MARK SCHEMES

(d) 1. Reference to nitrogen fixation ;

2. Reference to anaerobic process / eq ;

3. (Nitrogen fixed is) converted / synthesised to amino acids / proteins ;

4. Nitrogen combined with hydrogen ions (from carbohydrates) to give ammonium / ammonia (allow use of correct formulae) (also allow nitrogen reduced to form ammonia / ammonium) ;

5. Reference to nitrogenase ;

6. Ammonium / ammonia combined with glutamate / organic acid to give glutamine / amino acid ;   **[3]**

(Total 6 marks)

3 (a) Small intestine / ileum / duodenum ;   **[1]**

(b) Enables grip / eq ;

Prevents it being carried away by peristalsis / movement of food through intestine ;   **[2]**

(c) No mouth / alimentary canal / digestive system / eq ;

Lack of sense organs / sense receptors / reduced nervous system ;

Thick / enzyme resistant / tough, tegument / covering / eq / production of mucus, to protect against enzymes ;

Reference to ability to tolerate low oxygen conditions / anaerobic conditions ;

Reference to prolific reproductive capacity ;

Reference to being a hermaphrodite ;

Reference to flat body shape / large surface to volume ;   **[2]**

(d) *Rhizopus* is a saprobiont / eq ;
Taenia / parasites feed on host and *Rhizopus* feeds on dead material ;
No digestion / no digestive enzymes / food already digested in *Taenia* and external digestion / eq in *Rhizopus* ;
Absorption of food over whole body surface in *Taenia* and reference to mycelium / hyphae in *Rhizopus* ;   **[3]**

(Total 8 marks)

Chapters 7, 8 and 9

1 (a) secondary consumer / top consumer / carnivore ;

primary consumer / herbivore ;   **[2]**

(b) reference to method of sampling / quadrats / eq ;

count plants ;

collect animals/organisms from given area / eq ;

identify all the animals found (in sample) ;

determine whether each is primary or secondary consumer / sort into trophic levels ;

count number of primary / secondary consumers (in sample) ;   **[4]**

(c) reference to energy loss ;

only small percentage / approx. 10–15%, of energy transferred between levels ;

reference to suitable reason for energy loss ;   **[2]**

(Total 8 marks)

2 (a) (taken up from atmosphere) 93 + 121 = 214 ;

(released into atmosphere) 90 + 120 + 1.6 + 0.1 + 5.5 = 217.2 ;

difference = 3.2 Gt (year^{-1}) ;   **[3]**

(b) planting of trees ;

reference to photosynthesis ;

carbon locked up in trees / reference to lignin or wood or cellulose ;

trees live for long time / hundreds of years ;

plants / trees die and become buried in swamps / eq ;

peat forms ;

fossil fuel forms ;

reference to shells / exoskeletons / limestone / calcareous rock ;

reference to carbon sink ;   **[4]**

(c) reference 93 Gt taken in, 90 Gt released / net of 3 Gt taken up / eq ;

some carbon, taken up by marine animals ;

forms calcium carbonate / forms shell / exoskeleton ;

example of marine animal (diatoms, corals, molluscs, etc) ;

used in photosynthesis ;

of algae / plankton ;

marine organisms die and form sediment on sea bed / settle on sea bed / eq ;

(sediments form) fossils / limestone / calcareous rock / chalk / oil / gas ;   **[4]**

(d) improved fuel efficiency / eq ;

increased use of public transport / eq ;

burn hydrogen / ethanol / gasohol / biogas (rather than petrol / oil) ;

produce / use (electricity) from nuclear / solar / wind / water / geothermal / wood / energy crop / straw / renewable energy ;

burning gas / methane in power station rather than oil / coal ;

reference to planting trees / forests / afforestation ;   **[3]**

(Total 14 marks)

3   (a) fuel / energy source / charcoal / burning for purpose ;

building material / reference to specific structural use / construction / furniture ;

paper making ;

turpentine / named chemical product from wood ;   **[Max 2 marks]**

(b) agriculture / cultivation / industrial use / urbanisation ;

reference to patterns showing planted crops / rows / roads / terrances / increased population / increased demand for food / economic reasons ;   **[2]**

(c) (biodiversity) reduced / fewer, species / varieties / types / eq.;

reference to loss of habitats / niche / disruption of food chains ;   **[2]**

(d) fewer, trees / plants, less photosynthesis ;

reference to absorption of carbon dioxide by, trees / plants / photosynthesis ;

burning (of trees) releases carbon dioxide ;

rotting / decay, (of felled trees) / respiration, releases carbon dioxide or soil disturbance releases carbon dioxide ;   **[Max 3 marks]**

(e) Using, trees / forest, in such a way as not to, destroy / reduce / use it up ;

Over a (relatively) long time period (minimum 5 years) ;

Timber is, cut / eq. at the same rate as it can be, grown / regenerated / replanted / eq.;

Example – e.g. coppicing / selective felling / rubber tapping / pollarding / replanting / reforestation ;   **[Max 3 marks]**

(Total 12 marks)

4   (a) 0, 0, 0, 0, 0, 10, 10, 25, 12, 6, 9, 6, 6, 4 ;

total = 88 ;   **[2]**

(b) site 1 score higher than site 2 ;

score for site 1 2.5 / nearly 3 times more than site 2 ;

site 1 less polluted than site 2 ;

site 1 score indicates moderate pollution, site 2 indicates very severe pollution ;

site 1 score indicates moderate water quality, site 2 indicates very poor water/river quality ;

greater species diversity / more invertebrate groups / more species in site 1 / reference to one specific named example (but must be comparative) ;

those with species index number 10 all found in site 1 / none found in site 2 / eq ;   **[4]**

(c) some invertebrates miss the net / too small to catch ;

some invertebrates may not be dislodged / disturbed / animals may bury themselves deeper in mud ;

some animals washed down from upstream ;

reference to non-standard technique (e.g. length of kicking) ;   **[2]**

MARK SCHEMES

(d) pollutants affect oxygen levels / environmental conditions ;

different tolerances to pollutants / oxygen / eq ;

(therefore) only live / survive in conditions which suit them ;

invertebrates easy to sample / count / catch ;

[2]

(e) reference to leaching / draining of fertiliser / nitrates / slurry / soil into river ;

(fertiliser) contains nitrates / phosphates ;

leads to, nutrient enrichment / eutrophication (of river / water) ;

reference to increased algal growth / bloom on surface of water ;

reference to blocking out light to submerged plants ;

causes reduced photosynthesis (in submerged plants) ;

algae / plants die and decompose ;

(so more) oxygen depletion due to bacteria / eq ;

pesticides / eq kill / toxic ;

[4]

(f) detergents / washing / cleaning materials (contain phosphates) ;

sewage (contains phosphates) ;

[2]

(Total 16 marks)

Index

Page references in *italics* refer to a table or an illustration.

INDEX

INDEX

INDEX

INDEX

INDEX

INDEX